OpenGL 简约笔记：
用C#学面向对象的OpenGL

祝 威 编著

北京航空航天大学出版社
BEIHANG UNIVERSITY PRESS

内 容 简 介

本书由浅入深地讲解 OpenGL 的概念和用法,通过一个个简单的实例使读者对各个知识点一目了然。书中包括渲染管道、着色器、缓存、纹理、矩阵、光照模型、阴影、帧缓存、拾取、文字、用户界面、体渲染、透明和通用计算等内容。读者掌握这些之后,就可以自由地设计和编写中等规模的三维图形程序,并且能够渲染百万数量级顶点的模型。本书相关内容将在 Github 上持续更新,读者可参阅更多资料。

本书适用于熟悉 C、C++、C♯ 或 Java 等任何面向对象编程语言的读者,本书会对这些相关的基础内容进行必要的介绍。读者只需认真实践,完全可以掌握本书内容。

图书在版编目(CIP)数据

OpenGL 简约笔记:用 C♯学面向对象的 OpenGL/祝威编著. -- 北京:北京航空航天大学出版社,2019.4
ISBN 978 - 7 - 5124 - 2735 - 8

Ⅰ. ①O… Ⅱ. ①祝… Ⅲ. ①图形软件 Ⅳ.
①TP391.412

中国版本图书馆 CIP 数据核字(2018)第 266706 号

版权所有,侵权必究。

OpenGL 简约笔记:用 C♯学面向对象的 OpenGL
编　著　祝　威
责任编辑　剧艳婕

*

北京航空航天大学出版社出版发行

北京市海淀区学院路 37 号(邮编 100191)　http://www.buaapress.com.cn
发行部电话:(010)82317024　传真:(010)82328026
读者信箱:emsbook@buaacm.com.cn　邮购电话:(010)82316936
涿州市新华有限公司印装　各地书店经销

*

开本:710×1 000　1/16　印张:20　字数:426 千字
2019 年 4 月第 1 版　2019 年 4 月第 1 次印刷　印数:3 000 册
ISBN 978 - 7 - 5124 - 2735 - 8　定价:69.00 元

若本书有倒页、脱页、缺页等印装质量问题,请与本社发行部联系调换。联系电话:(010)82317024

前　言

本书将以 C♯ 为工具介绍 OpenGL 的相关知识。如果读者不太了解 C♯ 或面向对象程序设计，可参考本书最后的附录。

本书由浅入深地讲解 OpenGL 的概念和用法，通过一个个简单的实例让读者一目了然地认识 OpenGL 的各个知识点。本书包括渲染管道（Pipeline）、着色器（Shader）、缓存（Buffer）、纹理（Texture）、矩阵（Matrix）、光照模型（Lighting）、阴影（Shadow）、帧缓存（Framebuffer）、拾取（Picking）、文字（Text）、用户界面（UI）、体渲染（Volume Rendering）、透明（Transparency）和通用计算（Compute Shader）等内容。读者掌握这些内容后，就可以自由设计、编写中等规模的三维图形程序了，并且能够渲染百万数量级顶点的模型。

本书的网络资料包含了书中所有的完整代码，读者可以从北航出版社网站（www.buaapress.com.cn）的"下载专区"的相关页面获取。另外，读者也可以在 CSharpGL 托管页面（https://github.com/bitzhuwei/CSharpGL）找到本书的全部代码。如果读者对本书内容有任何疑问、指错或想进一步深入讨论，还可以在托管页面进行提问、讨论。本书最后的附录也提供了简单的 Github 入门教程。

感谢北京航空航天大学出版社提供的机会与平台，感谢家人以及业界的朋友在本书编写的过程中给予的支持和帮助，这里还要特别感谢编辑剧艳婕在本书出版的过程中提出的宝贵意见。希望本书能够帮助读者打开计算机图形学的大门。

作　者

2019 年 1 月

目 录

第 1 章 Hello OpenGL ·········· 1
1.1 从这里开始认识 ·········· 1
1.2 OpenGL 是什么 ·········· 1
1.3 如何使用 OpenGL ·········· 2
1.4 Hello OpenGL ·········· 6
1.5 辅助工具 ·········· 22
1.6 不含位置属性的顶点 ·········· 29
1.7 总 结 ·········· 31
1.8 问 题 ·········· 31

第 2 章 纹 理 ·········· 33
2.1 二维纹理 ·········· 33
2.2 其他类型的纹理 ·········· 40
2.3 多个纹理 ·········· 43
2.4 多个渲染方法 ·········· 46
2.5 总 结 ·········· 48
2.6 问 题 ·········· 48

第 3 章 空间和矩阵 ·········· 50
3.1 如何理解矩阵 ·········· 50
3.2 空间和矩阵的关系 ·········· 52
3.3 Pipeline 中的空间 ·········· 60
3.4 实例化渲染 ·········· 61
3.5 总 结 ·········· 63
3.6 问 题 ·········· 63

第 4 章 几何着色器 ... 64

- 4.1 介 绍 ... 64
- 4.2 示例:渲染法线 ... 66
- 4.3 总 结 ... 69
- 4.4 问 题 ... 69

第 5 章 光 照 ... 70

- 5.1 Blinn-Phong 光照模型 ... 70
- 5.2 光 源 ... 70
- 5.3 反射光 ... 70
- 5.4 Blinn-Phong 算法 ... 75
- 5.5 同时使用多个光源 ... 77
- 5.6 阴 影 ... 80
- 5.7 凹凸映射 ... 80
- 5.8 噪 声 ... 84
- 5.9 总 结 ... 88
- 5.10 问 题 ... 88

第 6 章 帧缓存 ... 89

- 6.1 名词术语 ... 90
- 6.2 附着点 ... 90
- 6.3 将 Texture 附着到 Framebuffer ... 91
- 6.4 附着 Renderbuffer ... 95
- 6.5 示例:Render to Texture ... 95
- 6.6 总 结 ... 99
- 6.7 问 题 ... 100

第 7 章 阴 影 ... 101

- 7.1 Shadow Mapping ... 101
- 7.2 Shadow Volume ... 110
- 7.3 模板缓存的初始化 ... 122
- 7.4 多光源下的阴影 ... 123
- 7.5 总 结 ... 124
- 7.6 问 题 ... 124

第 8 章 拾 取 ... 125

- 8.1 基 础 ... 125

8.2　在 DrawArraysCmd 命令下的拾取 ··· 140
8.3　在 DrawElementsCmd 命令下的拾取 ·· 146
8.4　拖拽顶点 ·· 157
8.5　总　结 ·· 162
8.6　问　题 ·· 163

第 9 章　文　字 ·· 164

9.1　固定尺寸且始终面向 Camera ·· 164
9.2　字形信息 ·· 169
9.3　三维世界的文字 ·· 174
9.4　总　结 ·· 177
9.5　问　题 ·· 177

第 10 章　简单的用户界面 ·· 178

10.1　指定区域的贴图 ·· 178
10.2　控件的布局机制 ·· 179
10.3　控件的事件机制 ·· 188
10.4　CtrlImage ··· 193
10.5　CtrlLabel ·· 196
10.6　CtrlButton ·· 200
10.7　总　结 ·· 202
10.8　问　题 ·· 202

第 11 章　轨迹球 ·· 203

11.1　轨迹球 ·· 203
11.2　使　用 ·· 204
11.3　设　计 ·· 205
11.4　四元数 ·· 209
11.5　总　结 ·· 210
11.6　问　题 ·· 210

第 12 章　体渲染 ·· 211

12.1　什么是体渲染 ·· 211
12.2　静态切片 ·· 212
12.3　动态切片 ·· 215
12.4　Raycasting ·· 217
12.5　总　结 ·· 223
12.6　问　题 ·· 224

第 13 章　半透明渲染 …………………………………………………… 225
13.1　跳跃着色法 ……………………………………………………… 225
13.2　与顺序无关的半透明渲染 ………………………………………… 226
13.3　Front to Back Peeling ………………………………………… 235
13.4　总　结 …………………………………………………………… 240
13.5　问　题 …………………………………………………………… 240

第 14 章　Transform Feedback Object ………………………………… 241
14.1　Transform Feedback 如何工作 …………………………………… 241
14.2　使用 Transform Feedback Object ………………………………… 244
14.3　轮流更新 ………………………………………………………… 245
14.4　粒子系统 ………………………………………………………… 251
14.5　总　结 …………………………………………………………… 254
14.6　问　题 …………………………………………………………… 254

第 15 章　Compute Shader ……………………………………………… 255
15.1　Compute Shader 简介 …………………………………………… 255
15.2　图像处理 ………………………………………………………… 257
15.3　粒子系统 ………………………………………………………… 261
15.4　总　结 …………………………………………………………… 262
15.5　问　题 …………………………………………………………… 263

附　录 ……………………………………………………………………… 264
附录 A　Github 入门 …………………………………………………… 264
附录 B　C#和面向对象入门 …………………………………………… 277
附录 C　解析简单的 wavefront(＊.obj)文件格式 ……………………… 298
附录 D　自制体数据的 2 种方法 ………………………………………… 302

第 1 章

Hello OpenGL

学完本章之后，读者将：
- 了解 OpenGL 是什么；
- 初次理解 OpenGL Pipeline 的工作流程；
- 使用 C♯ 编写简单的 OpenGL 程序。

1.1 从这里开始认识

　　大部分讲解 OpenGL 的书都是用 C\C++ 编写代码的，但本书全部使用 C♯ 编写 OpenGL 程序。这是因为，与 C\C++ 等语言相比，C♯ 需要的配置最少，运行环境最简单，调试环境最方便，最适合用来学习复杂的 OpenGL。

　　为了深入理解 OpenGL，尽可能发挥它的强大功能，本书使用了面向对象的编码方式，并涉及了一些数学和算法知识，因此本书要求读者具有普通程序员的编码技能。当然，本书会对相关的基础内容进行必要的介绍。读者只需认真实践，完全能够掌握本书内容。

　　学习一项技能，应该循序渐进、由浅入深。最初对 OpenGL 的认识，必然是不全面的，这很正常。因此，本书在对 OpenGL 最初的介绍中，会使用一些不全面甚至"不正确"的说法。这有利于读者逐步理解相关概念，降低学习难度。

　　无论是阅读本书还是代码，都应像剥洋葱一样，一层一层剥开，通读一遍，了解一些内容；再通读一遍，了解更多内容，直到彻底理解。

　　本书在中文段落中的代码一般会用 黑框 包围起来。但下列情况不会这样强调：
- 首次出现后又紧接着多次出现时；
- 要强调其他代码时。

1.2 OpenGL 是什么

　　用 OpenGL 可以编制多种三维图形程序（如 3D 游戏、工程仿真、石油勘探、影视特效、科研教学等)。那 OpenGL 具体是什么呢?

从表面看,OpenGL 就是一系列函数声明(function declaration),合理地组织这些函数,就可以在某种画布上渲染出图形。例如,用 C 或 C♯语言描述 OpenGL 中用于清空画布的函数 glClear(..),如下:

```
// clear buffers to preset values
// mask: Bitwise OR of masks that indicate the buffers to be cleared
//The three masks are
//GL_COLOR_BUFFER_BIT, GL_DEPTH_BUFFER_BIT, and GL_STENCIL_BUFFER_BIT
void glClear(uint mask);
```

注意:OpenGL 只有函数的声明,而没有函数的定义(function definition)。也就是说,OpenGL 不负责实现这些函数(function implementation)。实现 OpenGL 给出的这些函数是显卡驱动的任务。当然,也可以自己编写一套 OpenGL 的实现,只是运行效率会比较低。

1.3 如何使用 OpenGL

1.3.1 准备工作

OpenGL 只有函数声明,显然不可能直接使用它。所谓的使用 OpenGL,意思是调用实现了 OpenGL 的显卡驱动程序。在能够正常地使用 OpenGL 绘制三维图形前,程序员需要做一些准备工作。

● 首先要找到 OpenGL 函数。

在 Windows 下,找到 OpenGL 函数很简单,做一个 extern 的函数声明即可。例如 C♯语言的 glClear(..)的声明如下:

```
[DllImport("opengl32.dll", SetLastError = true)] // C♯ Attribute
static extern void glClear(uint mask);
```

但如果是扩展的 OpenGL 函数(例如 uint glCreateShader(uint shaderType);),则要使用动态加载的方式,从系统文件 opengl32.dll 中找到函数指针。

● 接着,创建 Render Context。

没有这个 Context,任何对 OpenGL 函数的调用都是无效的。创建 Context 的步骤很繁琐,其中任何一点出错,就可能产生各种奇怪的问题,而且难以查找。

● 然后准备模型。

为了便于学习,可以只用一个三角形当模型,但在学习光照等环节时,没有恰当的模型很难体会到算法的精髓。

● 最后，要先领会很多孤立的函数和矩阵变换。

把模型数据正确地传送到显卡，并渲染出来，中间涉及几十个函数、各种神似的参数和复杂的绑定关系。其中任何一点出错，都很难调试，让人心态崩溃。矩阵也是 OpenGL 里很难理解的一个工具。为了实现各种方位变换、光照效果，矩阵变换也是绕不开的问题。

综上所述，晦涩繁琐、易错难改是初学 OpenGL 的最大困难。为了解决这些问题，本书以 CSharpGL 为基础，讲解如何开始使用 OpenGL。其好处之一就是，上述准备工作都可以省了。

1.3.2　开源库

开源库（CSharpGL）是笔者设计编写维护的，它包含下述部分：

① CSharpGL：这是一个 OpenGL 库，封装了 OpenGL 的功能，编译出来的是一个库文件 CSharpGL.dll；

② CSharpGL.Windows：封装了在 Windows 下使用 OpenGL 的准备工作。编译出来的是一个库文件 CSharpGL.Windows.dll；

③ Infrastructures\：有一些辅助代码。既有必要出现，又不属于上述部分；

④ 其他：使用 CSharpGL 的示例集合。

有了 CSharpGL，读者只需做如下准备工作：

● 下载安装 Visual Studio 2013；

● 在本书网络资料中找到 CSharpGL.zip，用 Visual Studio 2013 打开 CSharpGL.sln 解决方案；

● 完毕。读者可以直接运行各个示例项目，看看 CSharpGL 都能做些什么。

1.3.3　渲染流程

OpenGL 应用程序开发者只需告知 OpenGL 想做些什么（渲染三维场景、执行并行计算等），而不需知道 OpenGL 如何去完成这些任务。OpenGL 会在一系列复杂的处理之后完成这些任务，这一系列复杂的处理过程就是 OpenGL 的渲染流程（Pipeline）。这类似《植物大战僵尸》这样的塔防游戏，玩家只需放置需要的植物（坚果墙、向日葵、豌豆炮、南瓜、磁力菇等），它们就会自动发挥自己的作用，玩家不必关心豌豆是如何神奇地修炼成精并发出炮弹的。

这里来初次认识一下简化的 Pipeline。OpenGL 处理模型数据主要分 3 步，如图 1-1 所示。

1. 准备数据

三维世界里的物体，可以由组成物体的点以及点之间的连线（直线）来描述。例如图 1-2 就用点和连线描述了一个立方体和一个球体。

一个点最重要的数据就是其位置和颜色，该点就是 OpenGL 中的顶点（Ver-

图 1-1　简化的 OpenGL Pipeline

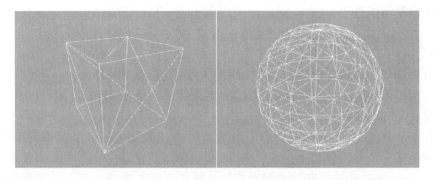

图 1-2　用点和连线描述的立方体和球体

tex)。用这种方式描述的物体称为一个模型(Model)。

　　OpenGL 处理的模型，是由一个个顶点组成的。对于每个顶点，都可能需要指定它的位置、颜色、法线和纹理坐标等属性(Attribute)。顶点的位置属性一般必须指定，而其他属性则未必。但是从程序的角度看，位置属性和其他属性是并列关系。

　　为了便于解释，这里以一个立方体 Cube 模型为例。立方体具有很特殊的性质，以后也将用它学习观察 OpenGL 的各种功能用法。CSharpGL 中用图 1-3 所示的注释来记录立方体的顶点顺序和坐标轴。这样的方式既直观，又不像图片那样占用大量的空间。

　　根据图 1-3，首先指定立方体 8 个顶点的位置属性：

```
private const float halfLength = 0.5f;
private static readonly vec3[] positions = new vec3[]
{
    new vec3(+halfLength, +halfLength, +halfLength),  // 0
    new vec3(+halfLength, +halfLength, -halfLength),  // 1
    new vec3(+halfLength, -halfLength, +halfLength),  // 2
    new vec3(+halfLength, -halfLength, -halfLength),  // 3
    new vec3(-halfLength, +halfLength, +halfLength),  // 4
    new vec3(-halfLength, +halfLength, -halfLength),  // 5
    new vec3(-halfLength, -halfLength, +halfLength),  // 6
    new vec3(-halfLength, -halfLength, -halfLength),  // 7
};
```

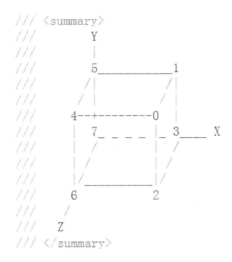

图 1-3 用字符画描述 Cube

这里只指定位置属性,其他属性暂不涉及。

有了位置,下面要指定这些顶点的连接关系。立方体有 6 个面,用 12 个三角形来描述这 6 个面,每个三角形需要指定 3 个顶点。用一个 uint[]类型的数组 indexes 指定 12 个三角形,代码如下:

```
private static readonly uint[] indexes = new uint[]
{
    0, 2, 1,  1, 2, 3, // +X faces.
    0, 1, 5,  0, 5, 4, // +Y faces.
    0, 4, 2,  2, 4, 6, // +Z faces.
    7, 6, 4,  7, 4, 5, // -X faces.
    7, 5, 3,  3, 5, 1, // -Z faces.
    7, 3, 2,  7, 2, 6, // -Y faces.
};
```

上述代码中,每 3 个数值描述一个三角形,每个数值指定此三角形的顶点在 positions 里的索引。例如, indexes[0] 、 indexes[1] 、 indexes[2] 指定了第一个三角形,其顶点位置分别为 positions[indexes[0]] 、 positions[indexes[1]] 和 positions[indexes[2]] 。

2. 顶点着色器

着色器(Shader)是一段负责给图形上色(或为上色做准备)的程序,其形式类似于 C 语言。OpenGL 开始渲染后,首先会对每个顶点都执行一个简短的程序,这个程序就叫 Vertex Shader。

Vertex Shader 有 2 个用处：

> 指定顶点在 Clip Space 里的位置（即设置内置变量 vec4 gl_Position; 的值）。
> Clip Space 的相关内容将在矩阵章节具体介绍。现在只需知道，为立方体指定的顶点位置是在三维空间里的，而这最终要画到二维的窗口上，所以必然要进行坐标变换。坐标变换就是矩阵的功能。

> 向后续的 Fragment Shader 传递数据。为了实现各种光照效果，向 Fragment Shader 传递顶点的位置、颜色、纹理坐标等属性。

3. 线性插值

立方体只有 8 个顶点，然而 OpenGL 却要画出成片的三角形，这是通过 OpenGL 内部的插值功能实现的。在 Vertex Shader 完成后，OpenGL 会根据 indexes 生成所有的三角形，然后根据三角形的三个顶点的位置、颜色等属性，通过插值的方式，为三角形覆盖到的各个像素生成其位置、颜色等属性值。上文说明了 Vertex Shader 可以向 Fragment Shader 传递数据。这些数据，也将作为顶点的属性，一并进行插值。

具体的插值过程完全由 OpenGL 自动完成，OpenGL 应用程序不需也无法干预。

4. Fragment Shader

经过插值处理，画布上的一个像素所在的位置，每被三角形覆盖一次，就会产生一个富含顶点属性的片段（Fragment）。Fragment Shader 会对每个 Fragment 执行一次，其输出结果就是它所在像素的候选颜色值。

在 Fragment Shader 中，一个像素的颜色是由四维向量 vec4(r, g, b, a); 描述的。其中 r、g、b、a 的范围都是 [0，1]。如果 Shader 输出的值超出此范围，OpenGL 会自动将其裁切到范围内。

1.3.4 小　结

这是首次认识 Pipeline，因此省略了很多环节。真正完整的 Pipeline 要复杂得多，本书将逐步展开予以介绍。

Pipeline 是 OpenGL 的核心，是一个高度灵活的算法（可以自定义 Shader、控制开关）。充分利用 Pipeline，就可以实现光照、阴影、拾取、透明等各种效果和功能。

1.4　Hello OpenGL

本节一步步地介绍如何完成第一个 OpenGL 程序，画出一个立方体。读者可在本书网络资料上的 c01d00_Cube 项目中找到此示例的完整代码。建议读者先打开此项目，以逐步执行的方式调试运行几次，对项目代码建立一些感性的了解，再详细阅读下面的章节。请注意随时记得 Pipeline。OpenGL 所做的一切都是围绕 Pipeline 进行的。

Pipeline 需要三类信息：模型、Shader 和状态开关。最后调用 OpenGL 的渲染命令即可。状态开关用于控制 OpenGL 的各种状态，例如控制多边形渲染模式的 glPolygonMode(..)，控制线宽度的 glLineWidth(..) 等。

1.4.1 新建项目

首先，在本书网络资料中找到 CSharpGL.zip 进行解压缩，用 Visual Studio 打开 CSharpGL.sln 解决方案。将解决方案中的 Demos 文件夹和 OpenGLviaCSharp 文件夹下的所有项目都删除，以便从零开始创建新项目。然后，新建一个项目，如图 1-4 所示。

图 1-4 新建项目 c01d00_Cube

新建一个"Windows 窗体应用程序"，项目命名为"c01d00_Cube"。然后，为该项目引用必要的 DLL 库项目，如图 1-5 所示。

只需引用 CSharpGL.dll 和 CSharpGL.Windows.dll。CSharpGL.dll 封装了 OpenGL 函数，类似于在 C++ 中引用了 gl.h 头文件。CSharpGL.Windows.dll 封装了在 Windows 下使用 OpenGL 的初始化工作，类似于在 C++ 中引用了 glut.h 头文件。

OpenGL 简约笔记：用 C♯ 学面向对象的 OpenGL

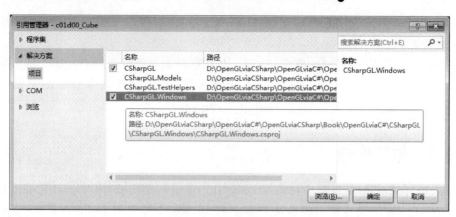

图 1-5 添加引用

1.4.2 使用 WinGLCanvas 控件

项目创建完成后，会自动提供一个 Form1 窗口。在这个窗口上添加一个 WinGLCanvas 控件。WinGLCanvas 控件有如下功能：

① OpenGL 渲染的所有内容都将显示在此控件范围内；

② 封装了 OpenGL 的初始化过程；

③ 提供渲染事件和两种渲染触发方式（自动或手动）。

用 Visual Studio 编译刚刚创建的项目，打开 Form1 窗口，就可以在工具箱里找到 WinGLCanvas 控件，如图 1-6 所示。

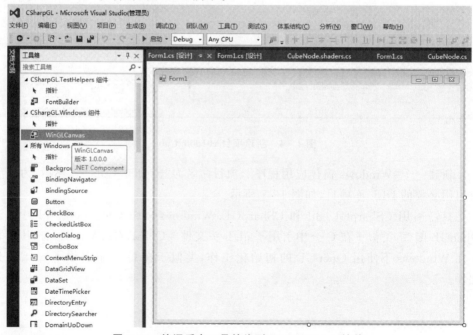

图 1-6 编译后在工具箱找到 WinGLCanvas 控件

从工具箱中拖拽一个 WinGLCanvas 控件到窗口 Form1 上，设置其 Dock 属性为 Fill，保存、编译，此时 Form1 里的控件是纯黑色的，如图 1-7 所示。

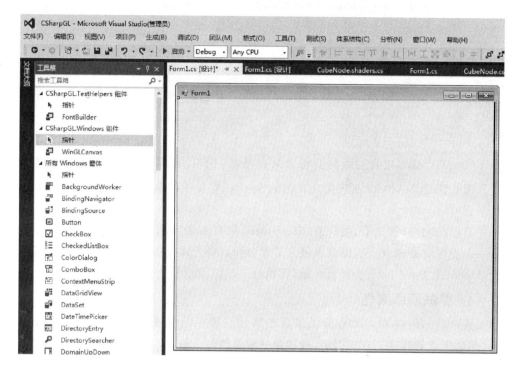

图 1-7　添加 WinGLCanvas 控件到窗口

1.4.3　模型数据 IBufferSource

项目、控件已经准备好，新建代码文件 CubeModel.cs，用以向 OpenGL 提供模型数据，代码如下：

```
namespace c01d00_Cube
{
    class CubeModel : IBufferSource
    {
        #region IBufferSource 成员

        public IEnumerable<VertexBuffer> GetVertexAttribute(string bufferName)
        {
            throw new NotImplementedException();
```

```csharp
            }

            public IEnumerable<IDrawCommand> GetDrawCommand()
            {
                throw new NotImplementedException();
            }

            #endregion
        }
    }
```

OpenGL能够接收的数据要符合某种格式,接口 IBufferSource 描述了这个格式。因此让CubeModel实现接口IBufferSource,它提供的数据就能够被OpenGL理解了。

目前,读者只需知道,接口IBufferSource只有两个方法,一个提供顶点的属性,另一个提供渲染命令。渲染命令包含了索引和渲染方式等信息。渲染命令每启动一次,OpenGL Pipeline就会运行一遍,即执行一次渲染过程。

1. 提供顶点属性

先来看如何向OpenGL提供顶点的属性。顶点的位置、颜色、法线、纹理坐标等,都是顶点的属性。在本例中,只涉及一种属性——位置。首先来看如何实现此方法,代码如下:

```csharp
class CubeModel : IBufferSource
{
    private static readonly vec3[] positions = ...

    public const string strPosition = "position";
    private VertexBuffer positionBuffer;

    #region IBufferSource 成员

    public IEnumerable<VertexBuffer> GetVertexAttribute(string bufferName)
    {
        if (strPosition == bufferName) // requiring position buffer.
        {
            if (this.positionBuffer == null)
            {
                // transform managed array to vertex buffer.
                this.positionBuffer = positions.GenVertexBuffer(
                    VBOConfig.Vec3, //mapping to 'in vec3 someVar;' in vertex shader.
```

```
                    BufferUsage.StaticDraw); // GL_STATIC_DRAW.
            }

            yield return this.positionBuffer;
        }
        else
        {
            throw new ArgumentException("bufferName");
        }
    }

    public IEnumerable<IDrawCommand> GetDrawCommand()
    {
        // ...
    }

    #endregion
}
```

此方法的返回类型是 IEnumerable<VertexBuffer>，实现时用的返回语句是 yield return。在开发实际应用程序时，可能会用到顶点数量很大的模型，这需要将一个顶点属性拆分为多段。用 yield return 就是为了方便开发者自行指定如何拆分。

目前读者可以直接将 IEnumerable<> 和 yield 忽略，更便于理解。忽略后，上述代码中的核心思想就是下面这两句代码：

```
    VertexBuffer positionBuffer = positions.GenVertexBuffer(VBOConfig.Vec3, BufferUsage.StaticDraw);
    return positionBuffer;
```

返回对象 positionBuffer 的类型是 VertexBuffer，可以将其视为一个空类型（void *）的数组，它与普通的 int[] 这样的数组的区别，就是它位于显卡内存中。

在 CSharpGL 中，一个 VertexBuffer 对象仅描述顶点的一种属性。上面 2 行代码的用处就是将 C# 托管内存中的数组 positions 上传到显卡内存，并返回生成的 VertexBuffer 对象。上传之后，C# 托管内存中的数组 positions 实际上就可以释放掉了。

如果读者接触过一些 OpenGL 知识，可能会联想到上面代码中的 GenVertexBuffer(..) 里使用了 OpenGL 里的下列函数：

```
//prepare n names for vertex buffers.
void glGenBuffers(int n, uint[] buffers);
// select buffer with specified target.
void glBindBuffer(uint target, uint buffer);
// setup data for buffer.
void glBufferData(uint target, int size, IntPtr data, uint usage);
```

稍作对比可发现,CSharpGL尽可能地减少了需要设置的参数,这有利于减少各种低级错误。

2. 提供渲染命令

渲染命令 IDrawCommand 是一个 interface 类型,它是对多种 OpenGL 渲染命令的抽象表示,例如:

```
void glDrawArrays(uint mode, int first, int count);
void glDrawArraysInstanced(uint mode, int first, int count, int primcount);
void glDrawElements(uint mode, int count, uint type, IntPtr indices);
void glDrawElementsInstanced(uint mode, int count, uint type, IntPtr indices, int primcount);
void glDrawElementsBaseVertex(uint mode, int count, uint type, IntPtr indices, int basevertex);
void glDrawElementsInstancedBaseVertex(uint mode, int count, uint type, IntPtr indices, int primcount, int basevertex);
...
```

在本例中,要使用的是 glDrawElements(..),它由 DrawElementsCmd 类型封装,代码如下:

```
class CubeModel : IBufferSource
{
    private static readonly uint[] indexes = ...
    private IDrawCommand drawCommand;

    #region IBufferSource 成员

    public IEnumerable<VertexBuffer> GetVertexAttribute(string bufferName)
    {
        ...
    }

    public IEnumerable<IDrawCommand> GetDrawCommand()
    {
        if (this.drawCommand == null)
```

```
                {
                    // transform managed array to index buffer.
                    IndexBuffer indexBuffer = indexes.GenIndexBuffer(BufferUsage.StaticDraw);
                    // draw GL_TRIANGLES indexed by indexBuffer.
                    // glDrawElements(GL_TRIANGLES, indexBuffer.Length, GL_UNSIGNED_INT, 0);
                    this.drawCommand = new DrawElementsCmd(indexBuffer, DrawMode.Triangles);
                }

                yield return this.drawCommand;
            }

        #endregion
    }
```

此方法的返回类型是 IEnumerable<IDrawCommand>，实现时用的返回语句用的是 yield return。这与上一节里的情形一致，所以读者仍可直接忽略 IEnumerable<> 和 yield。此方法的关键是类似的 3 行，如下所示：

```
IndexBuffer indexBuffer = indexes.GenIndexBuffer(BufferUsage.StaticDraw);
IDrawCommand drawCommand = new DrawElementsCmd(indexBuffer, DrawMode.Triangles);
return drawCommand;
```

GenIndexBuffer(..) 方法与 GenVertexBuffer(..) 类似，不再重复介绍。由此 3 行核心代码可见，将 C# 托管内存中的数组 indexes 上传到显卡内存，并记录生成 IndexBuffer 的对象，然后将其作为参数提供给渲染命令 DrawElementsCmd。上传之后，C# 托管内存中的数组 indexes 实际上就可以释放掉了。

1.4.4 Shader

根据 Pipeline 中的三个阶段准备好了模型数据，现在需要准备 Vertex Shader 和 Fragment Shader。OpenGL 用于编写 Shader 的语言被称为 GLSL。OpenGL 能够直接接收字符串形式的 Shader 源代码并进行编译。

1. Vertex Shader

Vertex Shader 会对每个顶点执行一次，本例中有 8 个顶点，应执行 8 次。本例使用的 Vertex Shader 代码如下：

```
// vertex shader.
#version 150 // GLSL version.
in vec3 inPosition; // element in vertex buffer. Vertex' position in model space.
uniform mat4 mvpMatrix;
void main() {
    // transform vertex' position from model space to clip space.
    gl_Position = mvpMatrix * vec4(inPosition, 1.0);
}
```

这个 Vertex Shader 十分简单,和 C 程序类似,有个入口函数 main()。本例的 main()中只有一行代码,其作用是给 gl_Position 赋值,这就是顶点在 Clip Space 里的位置。

代码中 inPosition 变量的值来自数组 positions 里的一个元素。关键字 in 决定了每次执行 Vertex Shader 时,OpenGL 会将不同的元素提供给 inPosition。inPosition 这个变量名可以自定义,但 gl_Position 是 Shader 里内置的变量名,不能修改。

代码中的 mvpMatrix 是一个四维矩阵类型(mat4),作用是将指定的坐标位置变换为 Clip Space 里的坐标位置。关键字 uniform 决定了每次执行 Vertex Shader 时,mvpMatrix 中的数据都是相同的。

2. Fragment Shader

Fragment Shader 会对每个片段(Fragment)执行一次。本例的模型中只有 8 个顶点,但是在线性插值阶段后,OpenGL 根据索引 indexes 生成了大量的 Fragment。每个 Fragment 都与模型中直接提供的 8 个顶点一样,包含了顶点的各项属性信息(本例中只有位置属性,在后续章节中的 Fragment 还会包含颜色、纹理坐标、法线等属性)。因此,可以认为模型(Model)通过少量的顶点和索引信息,间接地向 OpenGL 提供了大量密集的顶点信息。本例使用的 Fragment Shader 代码如下:

```
// fragment shader.
#version 150 // GLSL version.
uniform vec4 color = vec4(1, 0, 0, 1); // default: red color.
out vec4 outColor;

void main() {
    outColor = color; // fill the fragment with specified color.
}
```

Fragment Shader 也类似于 C 程序,有个入口函数 main()。本例的 main()中只有一行代码,即将用户定义的颜色 color 赋值给 Fragment Shader 的输出变量 outColor 。

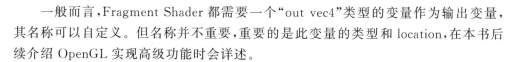

一般而言，Fragment Shader 都需要一个"out vec4"类型的变量作为输出变量，其名称可以自定义。但名称并不重要，重要的是此变量的类型和 location，在本书后续介绍 OpenGL 实现高级功能时会详述。

1.4.5 小 结

至此，OpenGL 所需的模型数据和 Shader 就齐全了，下面就可以调用 OpenGL 渲染图形了。不过这些内容显得有些支离破碎，难以将其关联起来，因此用下述伪代码描述这一切是如何合作的：

```
// Prepare data.
vec3[] positions = ... // VertexBuffer
uint[] indexes = ... // IndexBuffer

// set uniform variables in shaders.
ShaderProgram.mvpMatrix = ...
ShaderProgram.color = ...

// execute vertex shader for each vertex.
vec4[] gl_Positions = new vec4[positions.Length];
for (int i = 0; i < positions.Length; i++)
// for each element in VertexBuffer:
{
    var vs = new VertexShader();
    vs.inPosition = positions[i];
    vs.main(); // execute the vertex shader we provided.
    gl_Positions[i] = vs.gl_Position;
}

// linear interpolation.
Fragment[] fragments = LinearInterpolation(drawMode, indexes, gl_Positions);

Framebuffer fbo = ... // canvas container.

// execute fragment shader for each fragment.
for (int i = 0; i < fragments.Length; i++)
{
    var fs = new FragmentShader();
    fs.main(); // execute the fragment shader we provided.
    // assign fragment shader's output to framebuffer.
    fbo.colorAttachment.pixels[fs.FragCoord] = fs.outColor;
}

UseFramebuffer(fbo);
```

其中 Framebuffer 可以理解为一块画布,把 Fragment Shader 的输出结果赋值给 Framebuffer,就完成了 OpenGL 的渲染工作。尔后,开发者可以直接将 Framebuffer 的内容复制到窗口上,也可以留作他用(保存到文件、留作渲染阴影效果、支持拾取等)。

还可以发现,Shader 中的 uniform 变量类似 C♯ 中的静态变量,对于 Shader 的每次调用,其值都是相同的。

总览此伪代码,可以发现,开发者能够控制的有 3 部分:提供给 OpenGL 的模型数据、Vertex Shader 和 Fragment Shader。其他部分是 OpenGL 内置好且不能被修改的。可以将此伪代码描述的过程简单地理解为如图 1-8 所示的过程。

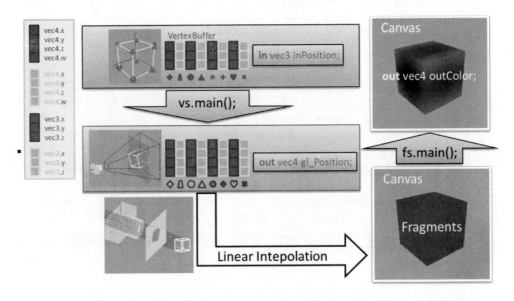

图 1-8 OpenGL 渲染流程图

下面要做的,就是将此伪代码实现为真正能够运行的代码。

1.4.6 场景和节点

本例中只绘制了一个 Cube,这个 Cube 可以视为一个节点(Node)。每次渲染时,都会为这个节点调用一次上一小节的伪代码,即启动一次渲染命令,执行一遍 Pipeline,让画布多一些内容。在实际的应用程序中可能会有多个节点,将这些节点链接成树结构并管理起来,就构成了一个场景(Scene)。例如图 1-9 展示的就是将在后续章节介绍的一个描述汉诺塔的场景。

此场景中被画出的节点有 18 个:地面、扁平的棱台、3 个柱子、10 个圆盘、红光源、绿光源和蓝光源。注意光源是用球体表示的。没有被画出的节点有 2 个:场景的根节点 GroupNode 和汉诺塔的根节点 GroupNode。左上方的 TreeView 控件描述

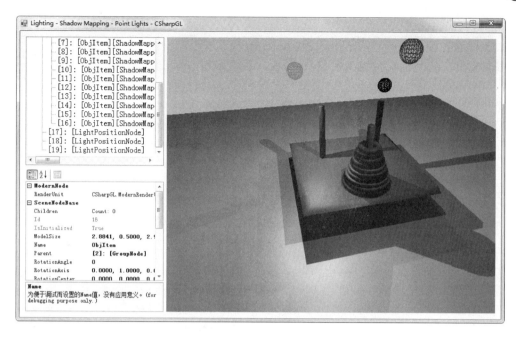

图 1-9 一个汉诺塔的场景

了这些节点之间的关系，左下方的 PropertyGrid 控件展示了其中一个节点的属性。树状的组织结构便于实现一些高级算法。

1.4.7 CubeNode

大多数情况下，一个 Node 由模型数据和至少一套 Shader（Vertex Shader + Fragment Shader）组成。CSharpGL 提供一个抽象类型 ModernNode，它封装了模型数据和 Shader 的初始化过程。为渲染 Cube 模型，创建 CubeNode 并继承它，代码如下：

```
namespace c01d00_Cube
{
    partial class CubeNode : ModernNode
    {
    }
}
```

1. 渲染功能

注意，ModernNode 只封装了初始化过程，并不提供渲染功能。一个原因是，并非所有节点都需要渲染，例如 GroupNode 的功能只是将场景的其他节点组织起来。另一个原因是，还有一些与渲染并列的功能，如果内置了渲染功能，就人为地特殊化了渲染功能。

因此手动添加接口 IRenderable。有了 IRenderable，自定义的 CubeNode 就能参加 CSharpGL 的统一渲染过程，这是面向对象编程的好处。代码如下：

```csharp
partial class CubeNode : ModernNode, IRenderable
{
    #region IRenderable 成员

    public void RenderBeforeChildren(RenderEventArgs arg)
    {
        // gets mvpMatrix
        ICamera camera = arg.Camera;
        mat4 projectionMat = camera.GetProjectionMatrix();
        mat4 viewMat = camera.GetViewMatrix();
        mat4 modelMat = this.GetModelMatrix();
        mat4 mvpMatrix = projectionMat * viewMat * modelMat;

        // a render unit wraps everything (model data, shaders, glswitches, etc.)
        //   for rendering
        ModernRenderUnit unit = this.RenderUnit;
        // gets render method
        // There could be more than 1 method(vertex shader + fragment shader)
        // to render the same model data. Thus we need a method array
        RenderMethod method = unit.Methods[0];
        // shader program wraps vertex shader and fragment shader
        ShaderProgram program = method.Program;
        //set value for 'uniform mat4 mvpMatrix;' in shader
        program.SetUniform("mvpMatrix", mvpMatrix);

        // render the cube model via OpenGL
        method.Render();
    }

    // ...

    #endregion
}
```

渲染功能在方法 RenderBeforeChildren(RenderEventArgs arg); 中得以实现。
需要注意以下几点：

➢ 矩阵 mvpMatrix 主要是借助参数 arg 获取的；

➤ 设置 uniform 变量值的方法 ShaderProgram. SetUniform<T>(string，T)；能接收各种 struct 类型的数据，要特别注意传入的 T 类型要与 Shader 里的 uniform 变量类型相同或匹配，否则会产生很难查找的错误。

2. 初始化数据

ModernNode 封装了数据的初始化过程，只需要将原始数据提供给它即可。代码如下：

```
partial class CubeNode : ModernNode, IRenderable
{
    public static CubeNode Create()
    {
        // model contains vertex buffer and index buffer.
        IBufferSource model = new CubeModel();
        // vertex shader and fragment shader.
        var vs = new VertexShader(vertexCode);
        var fs = new FragmentShader(fragmentCode);
        var array = new ShaderArray(vs, fs);
        // which attribute in shader maps to which vertex buffer.
        var map = new AttributeMap();
        map.Add("inPosition", CubeModel.strPosition);
        // build a render method.
        var builder = new RenderMethodBuilder(array, map);
        // create node.
        var node = new CubeNode(model, builder);
        // initialize node's data.
        node.Initialize();

        return node;
    }

    private CubeNode(IBufferSource model, params RenderMethodBuilder[] builders)
        : base(model, builders)
    {
    }
}
```

将 CubeNode 构造函数指定为 private ，禁止外部使用；提供静态的 Create() 方法来创建 CubeNode 对象。

在 Create() 中，读者已经认识了 IBufferSource、Vertex Shader 和 Fragment

Shader。这里重点说明 AttributeMap，它记录了 IBufferSource 里 GetVertexAttribute(string bufferName); 的参数 bufferName 和 Vertex Shader 里的 in 变量的对应关系。在本例中，只有 1 个 bufferName（即描述 positionBuffer 的 string strPosition = "position";）和 1 个 in 变量（即 Vertex Shader 中的 in vec3 inPosition;）。它们的对应关系通过 map.Add("inPosition",CubeModel.strPosition); 记录下来，用于 ModernNode 的初始化过程。

1.4.8 布置场景

节点已经准备完毕，下面把场景布置好，把节点放进去，就大功告成了。

1. 场 景

从三维世界的模型到二维画布的像素，是从某个角度观察的结果。因此要指定"从哪里看""看向哪里""何种倾斜角度"等参数。这些参数组织起来，就是摄像机（Camera）。创建摄像机和 Cube 节点，并交由 Scene 对象管理。代码如下：

```
public partial class Form1 : Form
{
    private void Form1_Load(object sender, EventArgs e)
    {
        var position = new vec3(5, 3, 4) * 0.5f;
        var center = new vec3(0, 0, 0);
        var up = new vec3(0, 1, 0);
        var camera = new Camera(position, center, up,
            CameraType.Perspective,
            this.winGLCanvas1.Width, this.winGLCanvas1.Height);
        SceneNodeBase cubeNode = CubeNode.Create();
        var scene = new Scene(camera);
        scene.RootNode = cubeNode;
        ...
    }
}
```

2. 动 作

场景有了，还需要指定动作（Action）。每个动作都会遍历场景中的所有节点，对其执行某种操作，例如更新位置、渲染、投射阴影、排序和拾取等。本例中，只需更新位置（TransformAction）和渲染（RenderAction）两种动作。代码如下：

```csharp
public partial class Form1 : Form
{
    private ActionList actionList;

    private void Form1_Load(object sender, EventArgs e)
    {
        // ...

        var list = new ActionList();
        list.Add(new TransformAction(scene));
        list.Add(new RenderAction(scene));
        this.actionList = list;
    }
}
```

要实现各种复杂绚丽的效果,一般都需要用各种不同的方式多次渲染场景。CSharpGL 将要渲染的数据打包为节点,将对数据的不同操作打包为动作,使得数据与操作分离,具有很大的灵活性。

3. 渲 染

最后,只需在 WinGLCanvas 的 OpenGLDraw 事件中调用动作即可。动作列表 ActionList 每执行一遍,窗口中的场景就会被重绘一次。代码如下:

```csharp
public partial class Form1 : Form
{
    private ActionList actionList;

    public Form1()
    {
        InitializeComponent();

        // render event.
        this.winGLCanvas1.OpenGLDraw += winGLCanvas1_OpenGLDraw;
    }

    private void winGLCanvas1_OpenGLDraw(object sender, PaintEventArgs e)
    {
        ActionList list = this.actionList;
        if (list != null)
        {
            // clear canvas.
            vec4 clearColor = this.scene.ClearColor;
```

```
            glClearColor(clearColor.x, clearColor.y, clearColor.z, clearColor.w);
            glClear(GL_COLOR_BUFFER_BIT | GL_DEPTH_BUFFER_BIT | GL_STENCIL_BUFFER_BIT);

            Viewport viewport = Viewport.GetCurrent();
            list.Act(new ActionParams(viewport));
        }
    }
}
```

注意，每次重绘前，都要用 glClear(..) 清除之前渲染的内容，原理就是将窗口所有位置的颜色都设置为用 glClearColor(..) 指定的颜色。

4. 完成

本项目主要内容如上所述。本节只挑选了创建一个示例项目的要点进行介绍，没有提及的部分相对简单，读者只需稍作分析即可理解。

请读者按照本书网络资料中的项目，对照着编写一遍此项目。遇到不懂处，请返回看本章的介绍。经过一次完整的编码，读者将能够看到如图 1-10 所示的窗口。

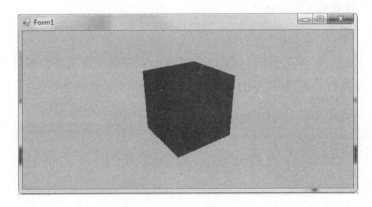

图 1-10　示例项目完成

1.5　辅助工具

这里向读者介绍 CSharpGL 提供的一个辅助工具，它能够显著提升做实验的效率。这个工具能够利用 C# 的[Attribute]特性和 PropertyGrid 控件，显示和修改节点的状态。为此新建项目 c01d01_PropertyGrid，它沿用 c01d00_Cube 的一切，只是在窗口左侧添加一个 TreeView 控件和一个 PropertyGrid 控件。

完成布置场景后，可以将场景里的节点与 TreeView 里的节点一一绑定。在 TreeView 里选择哪个节点，就会在 PropertyGrid 里显示哪个节点的属性。此代码十分简单，不再详述。此时编译运行项目，会看到如图 1-11 所示的窗口。

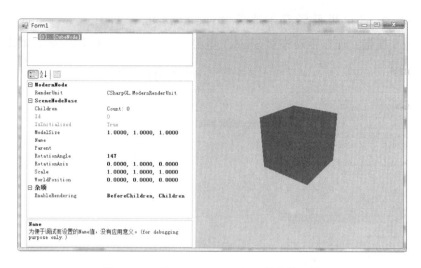

图 1-11　用 PropertyGrid 显示节点的属性

1.5.1　ModernNode 的属性

选中唯一的节点 CubeNode，PropertyGrid 会显示其属性。下面介绍其中常见的一些属性。

1. RenderUnit

RenderUnit 包含了渲染所需的主要数据（模型、Shader 和状态开关）。读者可以单击此属性右侧，打开观察其具体情况。

请读者沿着"CSharpGL.ModernRenderUnit"—"CSharpGL.RenderMethod"—"CSharpGL.ShaderProgram"—"UniformVariables"的顺序，打开 UniformVariable List Editor 窗口，如图 1-12 所示。

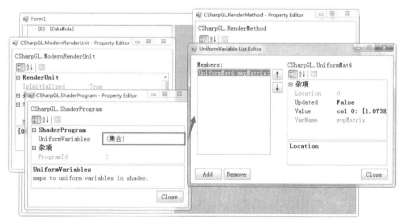

图 1-12　打开 Uniform Variable List Editor

可以看到，窗口里已经有了成员 mvpMatrix，它控制的是示例 Shader 里的 mvpMatrix 变量。现在手动添加 Shader 里另一个 uniform 变量 color ，如图 1-13 所示。

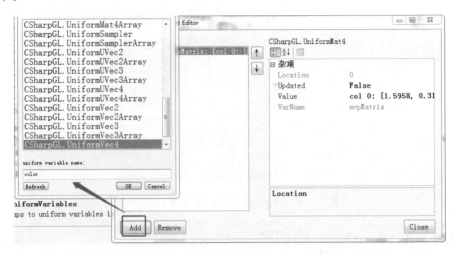

图 1-13　添加 uniform 变量 color

单击 OK 完成添加。下面就可以在右侧属性框里编辑颜色值了。尝试把 color 设置为绿色，如图 1-14 所示。

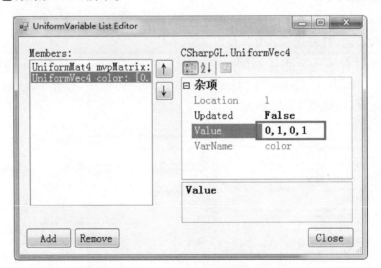

图 1-14　修改 color 属性值

注意，前三项分别表示 R、G、B 分量的值。最后一项应当为 1，具体原因请参考后续章节对矩阵的介绍。此时 Cube 的颜色立即随着指定的值改变了，如图 1-15 所示。

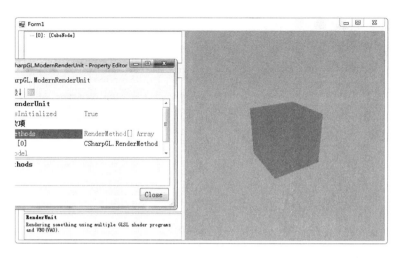

图 1-15 Cube 颜色随 uniform 变量改变

注意,由于在代码中每次渲染时都更新 mvpMatrix 的值,所以在此手动修改其值是无效的。

下面做一点儿有趣的事,并带出上文没有提及的重要对象——GLSwitch。它封装了对 OpenGL 各种状态的设置,例如启动或禁用 Blend、指定线条宽度等。

请读者沿着"CSharpGL.ModernRenderUnit"—"CSharpGL.RenderMethod"—"CSharpGL.ShaderProgram"—"SwitchList"的顺序,打开 IGLSwitch List Editor 窗口,如图 1-16 所示。

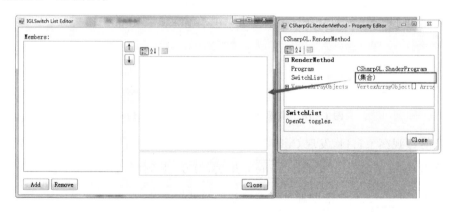

图 1-16 打开 IGLSwitch List Editor

该窗口可以让读者实时操纵节点里的 OpenGL 开关的状态。现在这个窗口是空的,即目前没有控制任何开关。添加一个 PolygonModeSwitch 开关,如图 1-17 所示。

单击 OK 按钮,窗口上就新增了一个控制 Polygon Mode 的开关。在窗口右侧,

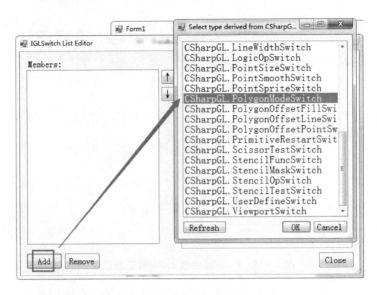

图 1-17 添加 PolygonModeSwitch

选择 Mode 属性的值为 Line ,如图 1-18 所示。

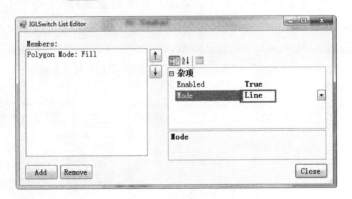

图 1-18 设置 PolygonMode 的属性值为 Line

此时再观察主窗口,发现 Cube 被用线框的方式画出来了,如图 1-19 所示。

这里使用的 PolygonModeSwitch 开关,实际上封装了 OpenGL 里的函数 glPolygonMode(..) 。请读者自行尝试其他类型的开关,观察效果。

2. 其他属性

Id 属性是生成节点对象时自动获得的编号,类似数据库中的主键,起到区分节点的作用。Name 属性不为 null 时,TreeView 控件会显示节点的 Name 值。

Children 和 Parent 分别指向此节点的子节点和父节点。本例中的 CubeNode 是 Scene 中的根节点,因此 Parent 属性值为 null。本例中的 CubeNode 没有子节点,因

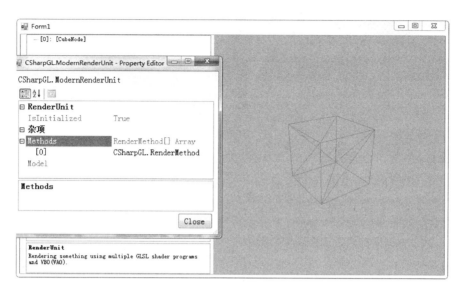

图 1-19　Cube 模型显示出线框形式

此 Children 的元素数量为 0。

IsInitialized 是只读属性，表明此节点是否已被初始化。一般在 PropertyGird 里看到的都是 true。ModelSize 描述此节点包含的模型大小，用 vec3 的三个分量 x、y、z 分别描述模型在 X 轴、Y 轴、Z 轴的长度。

RotationAngle 是模型的旋转角度，RotationAxis 是模型的旋转轴，Scale 是模型的缩放比例，WorldPosition 是模型的位置。这 4 项属性描述了模型在 World Space 里的方位。World Space 在后续的矩阵章节再做详细介绍。

EnableRendering 来自接口 IRenderable 的属性，是控制此节点是否参与渲染的开关。BeforeChildren 指调用 RenderBeforeChildren(RenderEventArgs); 方法；Children 指渲染此节点的子节点；AfterChildren 指调用 RenderAfterChildren(RenderEventArgs); 方法。在简单的情形里，一般设定此属性指为（BeforeChildren | Children），即先渲染此节点，再渲染其子节点。

1.5.2　自定义属性

下面，为 CubeNode 添加一个新属性，来体验一下使用 PropertyGrid 带来的便利。

上一节跋山涉水才找到控制 uniform 变量的窗口，这十分不便。这里就为 CubeNode 添加属性 CubeColor，从而直接在主窗口中控制 Cube 的颜色。首先在 CubeNode 类型里添加属性 CubeColor，代码如下：

```
private vec4 cubeColor = new vec4(1, 0, 0, 1);
[System.ComponentModel.Category("CubeNode")]
public System.Drawing.Color CubeColor
{
    get { return this.cubeColor.ToColor(); }
    set
    {
        vec3 v = value.ToVec3();
        this.cubeColor = new vec4(v.x, v.y, v.z, 1);
        ModernRenderUnit unit = this.RenderUnit;
        RenderMethod method = unit.Methods[0];
        ShaderProgram program = method.Program;
        program.SetUniform("color", this.cubeColor);
    }
}
```

注意，这里将 CubeColor 的类型设置为 System.Drawing.Color，这是因为.NET 平台已经为此类型在 PropertyGrid 里编写了十分直观的编辑面板。编译运行该项目，会在主窗口的 PropertyGrid 里看到新增的 CubeColor 属性，且修改属性值的方式十分直观，如图 1-20 所示。

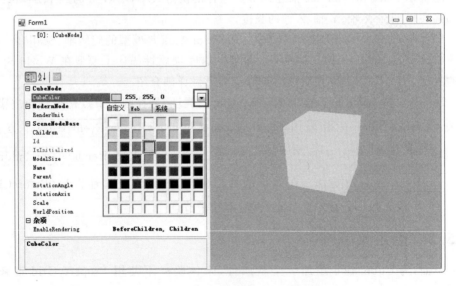

图 1-20 使用自定义属性 CubeColor

注意，CubeColor 属性指定了特性 [System.ComponentModel.Category("CubeNode")]，可以帮助开发者区分不同类型的属性。

1.5.3　开关列表窗口 IGLSwitch List Editor

在手动添加 GLSwitch 时，读者可以看到窗口中有很多选项。实际上读者可以随时编写新的 GLSwitch 类型，只要符合下述条件，新类型就会被自动添加到窗口中：

- 新类型继承自 GLSwitch 类型；
- 新类型有一个无参数的构造函数；

1.6　不含位置属性的顶点

在本章 1.3.3 小节中提到过下述一段内容：

OpenGL 处理的模型，是由一个个顶点组成的。对于每个顶点，都要指定它的位置、颜色、法线、纹理坐标等属性。顶点的位置属性一般都需要指定，而其他属性则未必。但是从程序的角度看，位置属性和其他属性都是并列关系。

那么真的存在不需要指定位置属性的模型吗？没有位置属性，如何渲染呢？本小节通过一个简单的示例说明如何在没有位置属性的情形下渲染图形。读者可在本书网络资料上的 Demos\ZeroAttributeInVertexShader 项目中找到此示例的完整代码。

不指定位置属性和其他任何属性的示例 Shader 代码如下：

```
// vertex shader.
#version 150 core
uniform mat4 mvpMat;
out vec2 passTexCoord;

void main()
{
    vec4 vertices[4] = vec4[4](vec4(-1.0, -1.0, 0.0, 1.0), vec4(1.0, -1.0, 0.0, 1.0), vec4(-1.0, 1.0, 0.0, 1.0), vec4(1.0, 1.0, 0.0, 1.0));
    vec2 texCoord[4] = vec2[4](vec2(0.0, 0.0), vec2(1.0, 0.0), vec2(0.0, 1.0), vec2(1.0, 1.0));

    passTexCoord = texCoord[gl_VertexID];

    gl_Position = mvpMat * vertices[gl_VertexID];
}

//-----------------------------------------------------------
```

```
// fragment shader.
#version 150 core

//uniform sampler2D tex;

in vec2 passTexCoord;
out vec4 outColor;

void main()
{
    outColor = vec4(passTexCoord, passTexCoord.x / 2 + passTexCoord.y / 2, 1.0f);
    //or: outColor = texture(tex, passTexCoord);
}
```

如上代码所示,虽然没有指定位置属性,但是在 Vertex Shader 内部镶嵌了 4 个顶点的位置和颜色信息。这样取巧的方式避开了位置属性的存在。这可能不具有大规模的实用性,但是说明了位置属性和其他属性是并列关系。

即使没有指定任何顶点属性,也要为此"模型"的 IBufferSource 接口提供完整的数据,代码如下:

```
// Model with zero vertex attribute.
public class ZeroAttributeModel : IBufferSource
{
    public IEnumerable<VertexBuffer> GetVertexAttribute(string bufferName)
    {
        throw new Exception("No vertex attribute buffer for this model!");
    }

    public IEnumerable<IDrawCommand> GetDrawCommand()
    {
        if (this.drawCmd == null)
        {
            int vertexCount = 4;
            this.drawCmd = new DrawArraysCmd(DrawMode.TriangleStrip, vertexCount);
        }

        yield return this.drawCmd;
    }

    private IDrawCommand drawCmd;
}
```

对此模型而言,并不存在顶点属性,因此 IBufferSrouce 的 GetVertexAttribute(..)方法永远不应被调用。此模型在 Shader 中镶嵌了 4 个顶点的属性,因此这里的渲染命令也应该指定 vertexCount 的值为 4。这样,上述 Vertex Shader 代码就会被执行 4 次。此示例渲染的结果如图 1-21 所示。

图 1-21 不含顶点属性的模型

1.7 总 结

本章揭开了编写 OpenGL 程序的第一层面纱,并提供了一个简单的辅助开发工具。

- OpenGL 是什么?

OpenGL 表现为一系列函数声明,一般由显卡驱动程序提供 OpenGL 的实现。

- OpenGL 渲染流程?

Prepare Data → Vertex Shader → Fragment Shader。

- 使用 CSharpGL 创建简单的 OpenGL 程序?

多练手,多实验,没有捷径。

1.8 问 题

带着问题实践是学习 OpenGL 最快的方式。这里给读者提出几个问题,作为抛砖引玉之用。

1. 本书网络资料中的 Cube 是在不停旋转的,这利用了一个 Timer 控件。请读者找到相关的代码,改变旋转轴或旋转速度,观察效果。

2. 画 Cube 略显无趣。请读者修改 CubeModel 中的 positions 和 indexes,尝试画出各种不同的模型。建议读者从只修改一个三角形开始,循序渐进,直至画出复杂有趣的图形。

3. 尝试使用不同的 GLSwitch，观察对 Cube 的影响。

4. 在 OpenGLDraw 事件的代码中，先用 ClearColor 和 Clear 函数清除窗口，再调用动作列表。请读者尝试新建一个动作类型，使其完成 ClearColor 和 Clear 的功能，然后将其加入当前的动作列表。这样就可以使事件代码在形式上十分整齐。

第 2 章

纹　理

学完本章之后,读者将:
- 了解纹理(Texture)是什么;
- 在 OpenGL 程序中使用二维纹理;
- 使用多个渲染方法。

2.1　二维纹理

把一张图片贴到三角形上,就可以让这个三角形面展现出这张图片的部分内容,这张图片就是一个二维纹理(2D Texture)。在第一章,Cube 被赋予了一个单独的纯色,并不能看出立体的效果。本章将把图 2-1(左)所示的纹理贴到模型的表面。

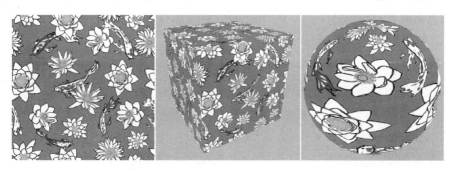

图 2-1　(左)纹理图片　(中)纹理贴到立方体模型　(右)纹理贴到球体模型

2.1.1　指定纹理坐标

纹理的哪个位置对应到模型的哪个顶点,是通过纹理坐标(Texture Coordinate)指定的。

本章将重新设计 Cube 的模型数据:指定 6 个面,每个面由 4 个顶点组成,所以共有 24 个顶点,并且指定顶点的纹理坐标属性。Cube 的顶点纹理坐标如下:

```csharp
private static readonly vec2[] uvs = new vec2[]
{
    new vec2(0, 0), new vec2(1, 0), new vec2(1, 1), new vec2(0, 1),
    new vec2(0, 0), new vec2(1, 0), new vec2(1, 1), new vec2(0, 1),
    new vec2(0, 0), new vec2(1, 0), new vec2(1, 1), new vec2(0, 1),
    new vec2(0, 0), new vec2(1, 0), new vec2(1, 1), new vec2(0, 1),
    new vec2(0, 0), new vec2(1, 0), new vec2(1, 1), new vec2(0, 1),
    new vec2(0, 0), new vec2(1, 0), new vec2(1, 1), new vec2(0, 1),
};
```

二维纹理坐标通常用 U 代替 X 轴，用 V 代替 Y 轴，原点在左下角，范围为[0，1]，如图 2-2 所示。

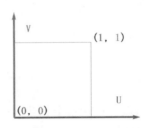

图 2-2　UV 坐标系

例如，对于一个记录了世界地图的纹理图片，可以将其贴到一个球体上，从而展现出立体的地球仪，如图 2-3 所示。

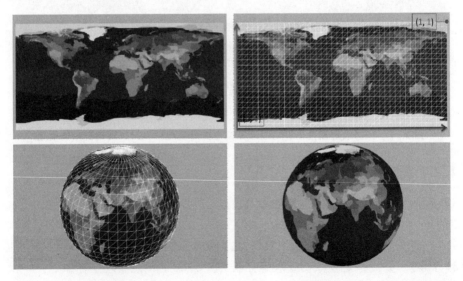

图 2-3　(左上)纹理内容　(右上)UV 对应顶点　(左下)顶点围成球体　(右下)最终效果

2.1.2 Vertex Shader

第一章的 Vertex Shader 只传递了系统变量 gl_Position 。这里也要把纹理坐标传递到 Fragment Shader,代码如下:

```glsl
// vertex shader.
#version 150 // GLSL version

in vec3 inPosition; // position.
in vec2 inTexCoord; // texture coordinate.

uniform mat4 mvpMatrix;

out vec2 passTexCoord; // texture coordinate passed to fragment shader.

void main() {
    gl_Position = mvpMatrix * vec4(inPosition, 1.0);
    passTexCoord = inTexCoord;
}
```

注意,顶点的位置是三维向量,顶点的纹理坐标是二维向量 vec2 。

2.1.3 Fragment Shader

在 Fragment Shader 中依据纹理坐标获取颜色,赋值给输出变量,代码如下:

```glsl
// fragment shader.
#version 150

// texture object.
uniform sampler2D tex;
in vec2 passTexCoord; // texture coordinate passed from vertex shader.
out vec4 outColor;

void main() {
    // assign texture's color at specified coordinate to output color.
    outColor = texture(tex, passTexCoord);
}
```

注意, texture(..) 是 GLSL 内置的函数,能够根据指定的坐标从指定的纹理中获取颜色值。

2.1.4 纹理对象

纹理对象是一种 OpenGL 对象，其创建方法之一是把客户端内存的图片信息上传到 GPU 端内存，同时记录此纹理对象的过滤器等各项信息。这里的客户端是指使用 OpenGL 的应用程序端，即开发者编写的应用程序，也可以称为 CPU 端。相应的服务器端是指在 GPU 上运行的显卡驱动程序，即执行 OpenGL 命令的程序，也可以称为 GPU 端。

"纹理对象"这个说法中的对象概念，既是 OpenGL 中的对象概念，也是面向对象编程中的对象概念。OpenGL 虽然是由一个个独立的函数组成的，但是 OpenGL 的设计理念完全符合面向对象的程序设计思想。OpenGL 中还有很多这样的对象，例如 Buffer 对象、Shader 对象、Shader Program 对象、Framebuffer 对象、Renderbuffer 对象、Transform Feedback 对象和 Query 对象等。从面向对象程序设计的角度看，这些对象都是某些类型(class)的实例。

上传二维纹理数据的 OpenGL 函数如下：

```
// upload image data to GPU side.
void glTexImage2D(uint target, int level, int internalFormat, int width, int height, int border, uint format, uint type, IntPtr data);
```

此函数的参数较多，不易理解。这里将其分为四类，分别介绍。

1. 内部格式

内部格式中，level、internalFormat 和 border 为一类，描述的是希望将纹理保存为何种格式。

一个 Texture 对象可能包含多个 Image。这些 Image 按从大到小、边长除以 2 的规则存在，例如由 6 个 level 的 Image 组成的一个 Texture 对象如图 2-4 所示。

图 2-4 由 6 个 level 组成的 Texture 对象

当模型被缩小到很小或者距离 Camera 很远时，贴在模型上的纹理就不需要很详细的数据了。此时 OpenGL 会用 level 值较大的那层 Image(即较小的图片)。

参数 `level` 指的是为哪个 level 设置 Image 数据。也就是说,调用一次 `glTexImage2D(..)`,只能指定其中一个 level 的 Image 数据,如果需要手动指定多个 level 的 Image 数据,就需要多次调用 glTexImage2D(..)。

参数 `internalFormat` 指的是要求 OpenGL 以何种格式保存贴图数据。本例中使用的是图片,一般使用 `GL_RGBA` 保存,这对应了 Shader 中的 `vec4` 保存颜色值的顺序。参数 `border` 指定纹理边框的宽度,只能为 0 或 1,一般设置为 0,即没有边框。

2. 外部格式

外部格式中,format、type 为二类,描述的是提供给此函数的贴图数据的格式。对于本例中的图片,其格式为 `GL_BGRA`。注意与内部格式 `GL_RGBA` 的细微差别。

3. 贴图数据

贴图数据中,data、width 和 height 为三类,描述的是贴图数据的指针和尺寸。在 C♯ 中,可以用如下代码获取和使用指针参数 data:

```
Bitmap bitmap = ...
BitmapData bits = bitmap.LockBits(..., PixelFormat.Format32bppArgb); // GL_BGRA
IntPtr data = bits.Scan0; // 获取参数 data
glTexImage2D(..., data); // 使用参数 data
bitmap.UnlockBits(bits);
```

4. 纹理目标

纹理目标 target 为四类,在本例中使用的是二维纹理,所以 target 设置为 `GL_TEXTURE_2D`。

综上所述,`glTexImage2D(..)` 的含义是,告诉 OpenGL 需要的纹理格式,提供贴图数据源及其格式,OpenGL 将上传数据并且自动转换格式。这符合 OpenGL 作为 API 的风格:开发者只需提出要求,OpenGL 负责实现。

2.1.5 过滤器

创建纹理对象时,还要为其设定过滤器。过滤器指定了对纹理进行采样时遇到各种情况的处理方式。同一纹理内容,在不同过滤器的作用下,会展现不同的效果。

1. Linear vs Nearest

在设定 GL_TEXTURE_MIN_FILTER 或 GL_TEXTURE_MAG_FILTER 属性时,如果设定的参数为 GL_LINEAR,那么 OpenGL 会用距离 UV 坐标最近的 4

个纹理值的加权平均值作为此处的颜色值,这会呈现出平滑的颜色。如果设定的参数为 GL_NEAREST,那么 OpenGL 会使用距离 UV 坐标最近的纹理值。两种采样方式的区别如图 2-5 所示。

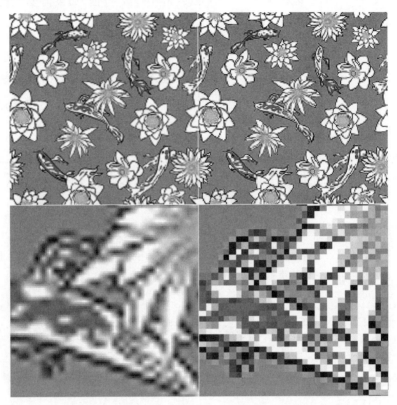

图 2-5　(左上)GL_LINEAR　(右上)GL_NEAREST
(左下)放大 GL_LINEAR　(右下)放大 GL_NEAREST

2. 超出边界的 Wrap 属性

在设定 Wrap 属性时,如果设定的参数为 GL_REPEAT,超出纹理范围的坐标整数部分会被忽略,形成重复效果;如果设定的参数为 GL_MIRRORED_REPEAT,超出纹理范围的坐标整数部分会被忽略,但当整数部分为奇数时进行取反,会形成镜像效果;如果设定的参数为 GL_CLAMP_TO_EDGE,超出纹理范围的坐标会被截取成 0 和 1,形成纹理边缘延伸的效果;如果设定的参数为 GL_CLAMP_TO_BORDER,超出纹理范围的部分会被设置为边缘色。上述参数各自的效果如图 2-6 所示。

注意,如果 Wrap 属性设置为 GL_CLAMP_TO_BORDER 时,边界颜色不一定必须是黑色,可以用 glTextureParameter(..)设置此时的边界颜色,代码如下:

```
float[] color = new float[] { 1.0f, 1.0f, 1.0f };
glTexParameterfv(GL_TEXTURE_2D, GL_TEXTURE_BORDER_COLOR, color);
```

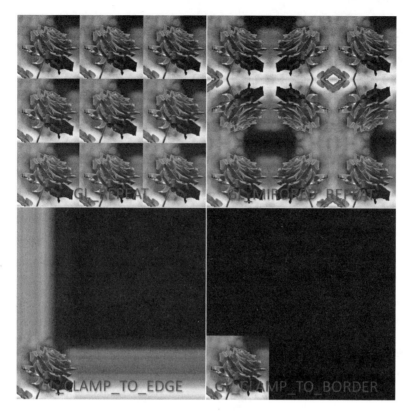

图 2-6 不同过滤器的效果

此代码会将纹理的边界颜色设置为白色。在 CSharpGL 中可以用 TexParameterfv 对象作为一个过滤器参数来指定边界颜色,代码如下:

```
Texture GetTexture()
{
    var bmp = new Bitmap("cloth.png");
    TexStorageBase storage = new TexImageBitmap(bmp, internalFormat: GL_RGBA, mipmapLevel: 1, border: true); // there is border.
    var texture = new Texture(storage,
        // border color: red(1, 0, 0).
        new TexParameterfv(GL_TEXTURE_BORDER_COLOR, 1, 0, 0),
        new TexParameteri(GL_TEXTURE_WRAP_S, GL_CLAMP_TO_BORDER),
        new TexParameteri(GL_TEXTURE_WRAP_T, GL_CLAMP_TO_BORDER),
        new TexParameteri(GL_TEXTURE_WRAP_R, GL_CLAMP_TO_BORDER),
        new TexParameteri(GL_TEXTURE_MIN_FILTER, GL_LINEAR),
        new TexParameteri(GL_TEXTURE_MAG_FILTER, GL_LINEAR));
    texture.Initialize();
    bmp.Dispose();
```

```
            return texture;
}
```

在 storage 对象中指定需要使用的 border(border = true),在过滤器列表中加入 GL_TEXTURE_BORDER_COLOR 类型的过滤器并指定颜色为红色(1,0,0)。为了让 border 生效,还要设置过滤器的 Wrap 模式为 GL_CLAMP_TO_BORDER。

2.1.6 其 他

上述模型和 Shader 都是数据,下面要把这些数据整合起来。在第一章 Node 的基础上,只需做下述几步。

首先,配置顶点属性的 AttributeMap,加入对纹理坐标属性的支持,代码如下:

```
var map = new AttributeMap();
// map.Add(varNameInShader, bufferNameInIBufferSource);
map.Add("inPosition", CubeModel.strPosition);
map.Add("inTexCoord", CubeModel.strTexCoord);
```

描述三维物体,一般都要有顶点的位置属性,这是顶点位置属性的特殊性。但从程序的角度看,顶点的位置、纹理坐标等属性,地位都是并列的。然后,指定纹理的 uniform 变量值,代码如下:

```
Texture tex = this.texture;
program.SetUniform("tex", tex);
```

可以看到,设置纹理的 uniform 变量与其他普通变量的形式是相同的。最后编译运行,如图 2-7 所示。

本例中,每个 Fragment 都与模型中直接提供的 8 个顶点一样,包含了顶点的位置属性和纹理坐标属性。这再一次证明,模型通过少量的顶点和索引信息,间接地向 OpenGL 提供大量而密集的顶点信息。

2.2 其他类型的纹理

2.2.1 一维纹理

一维纹理可以被视作高度为 1 的二维纹理。在 Fragment Shader 中使用一维纹理与使用二维纹理十分相似。不同点在于:在 Shader 中用 sampler1D 代替 sampler2D;在应用程序中用 glTexImage1D(..) 代替 glTexImage2D(..);纹理坐

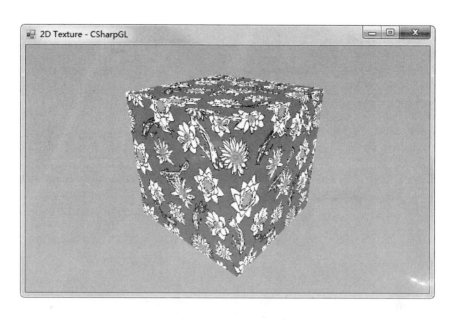

图 2-7　将图片贴到模型表面

标只需 X 坐标。

在绘制三维地形时,可以用不同的颜色表示海拔的高低。这时就可以用一维纹理对象将海拔值与颜色值对应起来。类似的,一维纹理也可以用于将密度、压力、流速等映射为颜色的情形。

2.2.2　二维数组纹理

一个应用程序中一般会有多个 Texture 对象。由于切换 Texture 对象需要消耗资源,OpenGL 提供了二维数组纹理(2D Array Texture)来减少切换。

在创建二维数组纹理对象时,需要将尺寸相同的多个图片按顺序提供给 OpenGL,使一个 Texture 对象实质上持有多个二维纹理对象的内容。在 Shader 中使用时,需要提供第三个坐标值(整数)来说明要使用哪个 Image。

注意,这与纹理的 level 是不同的。举例来说,一个由两张图片、level=4 构成的二维数组纹理对象,其内部数据如图 2-8 所示。

读者可以在本书网络资料上找到 Demos\Texture2DArray 示例项目,查看二维数组纹理的具体用法。本书将在后续章节介绍如何借助二维数组纹理来渲染文字。

OpenGL 也提供了一维数组纹理,其内部构成与二维数组纹理类似,只是每个 Image 都是一维的数据。

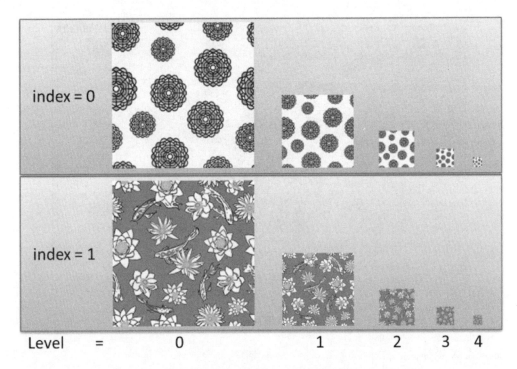

图 2-8 某个二维数组纹理对象的内容

2.2.3 三维纹理

三维纹理多用于体渲染。本书后续章节将介绍使用方法。

不存在"三维数组纹理"这样的类型。

2.2.4 Cubemap 纹理

Cubemap 纹理可以视作由 6 个二维纹理组成,且这 6 个纹理拼接成一个立方体。Cubemap 纹理在 Shader 中对应的类型为 samplerCube,其纹理坐标均在边长为 2 且中心位于原点的立方体表面,如图 2-9 所示。

Cubemap 纹理常用于制作天空盒,也可以用于环境映射,产生绚丽的效果。读者可以在本书网络资料上找到 Demos\EnvironmentMapping 项目,此项目既使用 Cubemap 创建了天空盒,又使用 Cubemap 为模型创建了反射和折射效果,如图 2-10 所示。

Cubemap 数组纹理(Cubemap Array Texture)是存在的,其类似于二维数组纹理,都是增加了一个索引坐标。

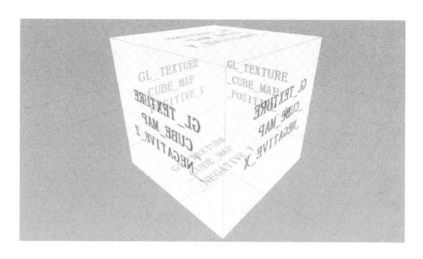

图 2-9 边长为 2 的立方体 Cubemap 纹理

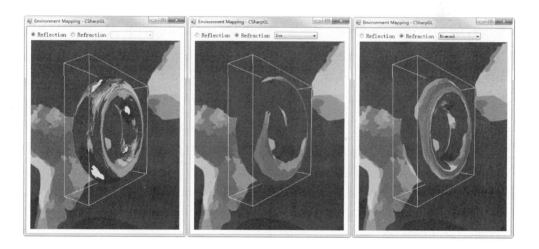

图 2-10 Cubemap 用于环境映射

2.3 多个纹理

  本节仍以 Cube 为例,说明如何同时使用多个二维纹理。最终实现的效果是让鱼群在花纹上游泳,如图 2-11 所示。

  模型数据不用修改,设置 Texture 的 uniform 变量值也是相同的,再多设置一个即可。不同的是,Fragment Shader 里要将两个 Texture 的颜色相加,代码如下:

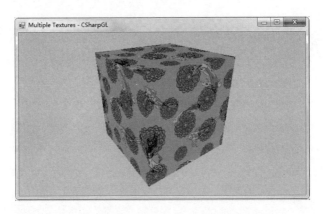

图 2-11　同时使用多个纹理的混合效果

```
// fragment shader.
#version 150

uniform sampler2D texture0;
uniform sampler2D texture1;

in vec2 passTexCoord;

out vec4 outColor;

void main() {
    vec4 c0 = texture(texture0, passTexCoord);
    vec4 c1 = texture(texture1, passTexCoord);

    outColor = vec4((c0 + c1).rgb / 2, 1);
}
```

除了简单相加的混合方式,读者可以发挥想象力,使用任何可能的计算方式进行混合,这就是 Shader 提供的巨大的灵活性。读者可在本书网络资料上的 c02d02_MultipleTextures 项目中找到此示例的完整代码。

OpenGL 内部提供了若干"晾衣架",每个晾衣架上对每种类型的 Texture 都有一个"挂钩",这样的晾衣架被称为纹理单元(Texture Unit)。第 n 个 Texture Unit 如图 2-12 所示。

当 n=0 时, glActiveTexture(GL_TEXTURE0); 会选中 OpenGL 的第 0 个 Texture Unit;之后对 Texture 的任何操作都是针对此 Texture Unit 下的 Texture。此时,可在此 Unit 下挂载不同类型的 Texture 对象各 1 个。挂载都是通过调用下述 OpenGL 函数完成的:

```
// bind specified texture to specified target of current selected Texture Unit.
void glBindTexture(uint target, uint texture);
```

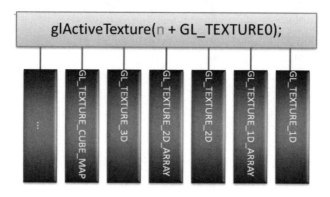

图 2-12 OpenGL 的第 n 个 Texture Unit

注意,将某个 target 类型的 texture 对象挂载(或者称为绑定)到一个 Texture Unit 后,之前被挂载到这一 target 上的 texture 对象就会被自动取消挂载。OpenGL 的 glActiveTexture(..)和 glBindTexture(..)联合起来操作 Texture 对象的伪代码如下:

```
// a TextureUnit object contains one texture for each type of target.
class TextureUnit
{
    uint target1D;
    uint target2D;
    uint target3D;
    // ...
}

TextureUnit[] textureUnits = new TextureUnit[GL_MAX_COMBINED_TEXTURE_IMAGE_UNITS];
uint currentTextureUnit = 0; // default current texture unit is 0.

// select current texture unit.
void glActiveTexture (uint textureUnit)
{
    currentTextureUnit = textureUnit - GL_TEXTURE0;
}

//currentTextureUnit → TextureUnit → target texture.
void glBindTexture (uint target, uint texture)
```

```
{
    TextureUnit unit = textureUnits[currentTextureUnit];
    switch (target)
    {
    case GL_TEXTURE_1D: unit.target1D = texture; break;
    case GL_TEXTURE_2D: unit.target2D = texture; break;
    case GL_TEXTURE_3D: unit.target3D = texture; break;
    // ...
    }
}
```

本例中,需要同时使用 2 个类型为 GL_TEXTURE_2D 的 Texture 对象。因此,必须在 2 个 Texture Unit 上分别挂载 1 个 Texture 对象。如果同时使用的 Texture 对象的类型不同,就可以挂载到同一个 Texture Unit 下。

在 CSharpGL 中,Texture 对象已经包含了自身的类型信息,只需指定将它挂载到哪个 Unit 上即可。代码如下:

```
texture0.TextureUnitIndex = 0; // glActiveTexture(0 + GL_TEXTURE0);
texture1.TextureUnitIndex = 1; // glActiveTexture(1 + GL_TEXTURE0);
```

实际上,给 Shader 的 uniform 变量类型(sampler1D、sampler2D 等)赋值时,赋予的是 Unit 的索引值,即图 2 - 12 中的 n。这样,Shader 就能够根据索引值和 uniform 变量类型(sampler1D、sampler2D 等),找到指定的 Texture 对象。这样就实现了同时使用多个 Texture 对象。

在不同的 OpenGL 实现里,支持的 Texture Unit 总数是不同的。OpenGL 规定最少应有 80 个,需要时可以用下述代码获取此数量:

```
var unitCount = new int[1];
glGetIntegerv(GL_MAX_COMBINED_TEXTURE_IMAGE_UNITS, unitCount);
```

类似于 Texture Unit,OpenGL 还存在其他的 Unit,这是 OpenGL 管理对象的一种方式。

2.4　多个渲染方法

实际应用中常常需要用多种方法渲染同一个模型。例如,高亮渲染、平时渲染,为了实现拾取、阴影而实施的渲染等。

CSharpGL 场景中的各个模型被组织为一个树结构,树结构的每个节点都支持多个渲染方法(Render Method)。一个渲染方法通常由 Shader、属性映射关系和若干 GLSwitch 组成。其中 Shader 是核心,其他内容都围绕 Shader 存在。

本节仍以 Cube 为例,为其实现 2 种渲染方法。一种渲染纯色的 Cube 线框,另一种渲染 2 个纹理叠加的颜色。这两种渲染方法都重复使用了之前介绍过的代码,现在只需在创建 Node 时将其一一提供给构造函数,代码如下:

```
public static CubeNode Create(Texture texture0, Texture texture1)
{
    IBufferSource model = new CubeModel(); // vertex buffer and index buffer.
    RenderMethodBuilder singleColorBuilder = ...; // to build render method [0]
    RenderMethodBuilder multiTextureBuilder = ...; // to build render method [1].
    // create node.
    var node = new CubeNode(model, singleColorBuilder, multiTextureBuilder);
    //...

    return node;
}

private CubeNode(IBufferSource model, params RenderMethodBuilder[] builders)
    : base(model, builders)
{ }
```

注意,在 ModernNode 的构造函数中,最后一个参数用 params 标记,意味着可以提供多个 RenderMethodBuilder 对象。在渲染时,根据配置选择使用其中一个渲染方法,代码如下:

```
public enum MethodType
{
    SingleColor = 0,
    MultiTexture = 1,
};

// Use which render method?
public MethodType CurrentMethod { get; set; }

// render this node.
public void RenderBeforeChildren(RenderEventArgs arg)
{
    //...
    // gets render method.
    RenderMethod method = this.RenderUnit.Methods[(int)this.CurrentMethod];
    //...
}
```

节点包含的各个 RenderMethod 会按照传入构造函数的 RenderMethodBuilder 的对象的顺序排列。图 2-13 所示的左右图分别展示了对同一个立方体模型使用不同的渲染方法得到的效果。

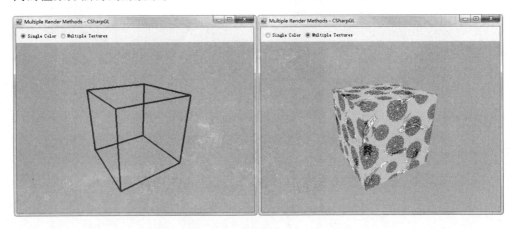

图 2-13 对同一模型使用不同的渲染方法

注意，左侧的渲染方法加入了 PolygonModeSwitch 和 LineWidthSwitch 两种 GLSwitch。这使得 OpenGL 以 Line 模式按指定宽度渲染 Cube，呈现出粗线框的样式。

读者可在本书网络资料上的 c02d03_MultipleRenderMethods 项目中找到此示例的完整代码。

2.5 总　结

- 二维纹理是什么？

把图片上传到显卡内存，"贴"到图形表面。

- 怎样在 OpenGL 程序中使用纹理？

通过 Texture Unit 的索引值和 Shader 中 uniform 变量的类型选中纹理。

- 可以使用多个渲染方法吗？

一个模型当然可以用多种方法渲染，合情合理。

2.6 问　题

带着问题实践是学习 OpenGL 的最快方式。这里给读者提出几个问题，作为抛砖引玉之用。

1. 尝试在 Vertex Shader 中使用纹理对象给顶点上色，然后将颜色传递到 Fragment Shader。观察效果，思考原因。

2. 用不同的图片代替本章使用的贴图,观察效果。

3. 手动为不同 level 指定内容不同的 Image,观察 OpenGL 在何种情形下使用的 level 值会发生改变。

4. 使用各种不同的算法(相加、相减、相乘等)来实施 2 个纹理的混合,观察效果。

第 3 章

空间和矩阵

学完本章之后,读者将能:
- 从方程组的角度理解矩阵;
- 理解 OpenGL Pipeline 中顶点的变换历程。

3.1 如何理解矩阵

3.1.1 方程组和矩阵

本章并不以绝对严格的数学定义和推导为主,而是从学习 OpenGL 的角度讲:矩阵是多元一次方程组的简便写法。例如,式(3-1)所示这样典型的二元一次方程组:

$$\begin{cases} 2x + 3y = 6 \\ 4x + 5y = 7 \end{cases} \quad (3-1)$$

若换用矩阵的形式书写,其形式如式(3-2)所示:

$$\begin{bmatrix} 2 & 3 \\ 4 & 5 \end{bmatrix} \begin{bmatrix} x \\ y \end{bmatrix} = \begin{bmatrix} 6 \\ 7 \end{bmatrix} \quad (3-2)$$

比较来看,矩阵形式的书写有这样的优点:

- ◆ 每个未知数 $\{x \ y\}$ 都出现且仅出现一次,这消除了公式中重复的内容。
- ◆ 系数与未知数被分开。方程组的本质特征存在于系数中,未知数仅仅起到占位符的作用。两者分开,利于突出本质。

这两种书写形式,可以看作是两种数的运算。

方程组里的数,称为实数;方程组由多个等式组成。矩阵形式里的数,称为"矩阵数"。方程组对应的矩阵形式由 1 个等式组成。

每个中括号里括起来的,就是一个"矩阵数"。式(3-2)描述的是一个等式,等式左侧有 2 个"矩阵数",即 $\begin{bmatrix} 2 & 3 \\ 4 & 5 \end{bmatrix}$ 和 $\begin{bmatrix} x \\ y \end{bmatrix}$,它们执行了一次"矩阵数"的乘法运算,等式右侧有 1 个"矩阵数",即 $\begin{bmatrix} 6 \\ 7 \end{bmatrix}$,是矩阵乘法运算的结果。

这样的矩阵运算能够存在的根据,就是因为这样的矩阵运算和方程组运算有一一对应的关系。虽然这样的"矩阵"数看起来不整齐,但是只将其视为方程组的另一个写法,是有道理的。

3.1.2 矩阵的乘法运算规则

两种形式相对照,可以从方程组的形式推理得到"矩阵数"的乘法运算法则。根据这一法则,可以解决渲染三维图形中的坐标变换问题。下面就以二元一次方程组为例,观察这一推理过程。可以将这一过程视为在二维世界里的坐标变换过程。

设有三组未知数 $\{x_1 \ y_1\}\{x_2 \ y_2\}$ 和 $\{x_3 \ y_3\}$。其中 $\{x_2 \ y_2\}$ 和 $\{x_3 \ y_3\}$ 有如式(3-3)所示的关系:

$$\begin{cases} a_{11}x_2 + a_{12}y_2 = x_3 \\ a_{21}x_2 + a_{22}y_2 = y_3 \end{cases} \leftrightarrow \begin{bmatrix} a_{11} & a_{12} \\ a_{21} & a_{22} \end{bmatrix} \begin{bmatrix} x_2 \\ y_2 \end{bmatrix} = \begin{bmatrix} x_3 \\ y_3 \end{bmatrix} \quad (3-3)$$

可以看到式(3-3)是对式(3-1)的泛化表示,左侧为方程组形式,右侧为等价的矩阵形式。其中 $\{x_1 \ y_1\}$ 和 $\{x_2 \ y_2\}$ 有如式(3-4)所示的关系:

$$\begin{cases} b_{11}x_1 + b_{12}y_1 = x_2 \\ b_{21}x_1 + b_{22}y_1 = y_2 \end{cases} \leftrightarrow \begin{bmatrix} b_{11} & b_{12} \\ b_{21} & b_{22} \end{bmatrix} \begin{bmatrix} x_1 \\ y_1 \end{bmatrix} = \begin{bmatrix} x_2 \\ y_2 \end{bmatrix} \quad (3-4)$$

然后,将式(3-4)中左侧的 $\{x_2 \ y_2\}$ 和右侧的 $\begin{bmatrix} x_2 \\ y_2 \end{bmatrix}$ 分别代入式(3-3)中,即可得到式(3-5):

$$\begin{cases} a_{11}(b_{11}x_1+b_{12}y_1)+a_{12}(b_{21}x_1+b_{22}y_1)=x_3 \\ a_{21}(b_{11}x_1+b_{12}y_1)+a_{22}(b_{21}x_1+b_{22}y_1)=y_3 \end{cases} \leftrightarrow \begin{bmatrix} a_{11} & a_{12} \\ a_{21} & a_{22} \end{bmatrix} \begin{bmatrix} b_{11} & b_{12} \\ b_{21} & b_{22} \end{bmatrix} \begin{bmatrix} x_1 \\ y_1 \end{bmatrix} = \begin{bmatrix} x_3 \\ y_3 \end{bmatrix}$$

$$(3-5)$$

将式(3-5)左侧的方程组整理一下,即可得到式(3-6):

$$\begin{cases} (a_{11}b_{11}+a_{12}b_{21})x_1+(a_{11}b_{12}+a_{12}b_{22})y_1=x_3 \\ (a_{21}b_{11}+a_{22}b_{21})x_1+(a_{21}b_{12}+a_{22}b_{22})y_1=y_3 \end{cases} \leftrightarrow \begin{bmatrix} a_{11} & a_{12} \\ a_{21} & a_{22} \end{bmatrix} \begin{bmatrix} b_{11} & b_{12} \\ b_{21} & b_{22} \end{bmatrix} \begin{bmatrix} x_1 \\ y_1 \end{bmatrix} = \begin{bmatrix} x_3 \\ y_3 \end{bmatrix}$$

$$(3-6)$$

将式(3-6)左侧的方程组,按照最初的方法用矩阵的形式书写,即可得到式(3-7):

$$\begin{bmatrix} a_{11}b_{11}+a_{12}b_{21} & a_{11}b_{12}+a_{12}b_{22} \\ a_{21}b_{11}+a_{22}b_{21} & a_{21}b_{12}+a_{22}b_{22} \end{bmatrix} \begin{bmatrix} x_1 \\ y_1 \end{bmatrix} = \begin{bmatrix} x_3 \\ y_3 \end{bmatrix} \leftrightarrow \begin{bmatrix} a_{11} & a_{12} \\ a_{21} & a_{22} \end{bmatrix} \begin{bmatrix} b_{11} & b_{12} \\ b_{21} & b_{22} \end{bmatrix} \begin{bmatrix} x_1 \\ y_1 \end{bmatrix} = \begin{bmatrix} x_3 \\ y_3 \end{bmatrix}$$

$$(3-7)$$

从式(3-7)可看到两个等式右侧均为 $\begin{bmatrix} x_3 \\ y_3 \end{bmatrix}$,左侧的 $\begin{bmatrix} x_1 \\ y_1 \end{bmatrix}$ 相同,那么可以推理出式(3-8):

$$\begin{bmatrix} a_{11}b_{11}+a_{12}b_{21} & a_{11}b_{12}+a_{12}b_{22} \\ a_{21}b_{11}+a_{22}b_{21} & a_{21}b_{12}+a_{22}b_{22} \end{bmatrix} = \begin{bmatrix} a_{11} & a_{12} \\ a_{21} & a_{22} \end{bmatrix} \begin{bmatrix} b_{11} & b_{12} \\ b_{21} & b_{22} \end{bmatrix} \quad (3-8)$$

从$\{x_1\ y_1\}$得到$\{x_2\ y_2\}$,再从$\{x_2\ y_2\}$得到$\{x_3\ y_3\}$,分别经历了$\begin{bmatrix} a_{11} & a_{12} \\ a_{21} & a_{22} \end{bmatrix}$和$\begin{bmatrix} b_{11} & b_{12} \\ b_{21} & b_{22} \end{bmatrix}$的运算。而从$\{x_1\ y_1\}$经历$\begin{bmatrix} a_{11}b_{11}+a_{12}b_{21} & a_{11}b_{12}+a_{12}b_{22} \\ a_{21}b_{11}+a_{22}b_{21} & a_{21}b_{12}+a_{22}b_{22} \end{bmatrix}$后,也可以得到$\{x_3\ y_3\}$,且只需一次计算。这就是矩阵乘法运算法则的原理。

3.1.3 矩阵乘法的意义

注意,这里读者应关注的不是严格意义的数学推理过程,而是推理过程中体现出来的矩阵乘法在描述三维世界时的意义。

在上一节的推理中,最初的假设是从$\{x_1\ y_1\}$得到$\{x_2\ y_2\}$,又从$\{x_2\ y_2\}$得到$\{x_3\ y_3\}$。将上一节的二元一次方程组替换为任意的N元一次方程组,其结论是不变的。如果$N=3$,即扩展到从$\{x_1\ y_1\ z_1\}$得到$\{x_2\ y_2\ z_2\}$,又从$\{x_2\ y_2\ z_2\}$得到$\{x_3\ y_3\ z_3\}$。这时,可以将$\{x_1\ y_1\ z_1\}$、$\{x_2\ y_2\ z_2\}$和$\{x_3\ y_3\ z_3\}$视为顶点的坐标,它描述的是OpenGL中这样的过程:把模型{在Model Space的坐标}变换为{在World Space的坐标},又把{在World Space的坐标}变换为{在View Space的坐标}。矩阵乘法能够完成这样的变换过程,而且还能将多个变换步骤合为1个,减少计算量。

矩阵就是一种简便写法,这种写法能够突出方程组的本质特征。在实际应用中使用矩阵时,直接省略对未知数$\{x_t\ y_t\ z_t\}$的书写。读者在看到矩阵时,应在脑海中将其扩充改写为方程组的形式,这样即可理解矩阵的作用。

3.2 空间和矩阵的关系

通过矩阵运算能够把顶点在一个空间中的坐标变换为在另一个空间中的坐标。在三维空间中,这些矩阵一般都是4×4的矩阵,用$\{x_t\ y_t\ z_t\ w_t\}$来描述顶点坐标(其中w_t往往是1)。本节从Model Space开始,介绍模型在各个空间的变换过程。

3.2.1 Model Space→World Space

如三角形、立方体这样简单的模型,可以直接在草稿纸上设计,复杂的模型如游戏人物,就需要在专门的设计软件上完成。但无论复杂程度如何,3D模型在设计时所在的空间都被称为Model Space(或者Object Space),模型的坐标都是相对于Model Space的原点来描述的。

设计好的模型被放到场景时,可能需要旋转、缩放或平移。这些操作都体现在World Space中,即World Space描述的是模型在场景中的方位。将模型从Model Space变换到World Space的矩阵称为Model Matrix,如图3-1所示。

如图3-1所示,在World Space中使用了2个Cube,它们分别被旋转、缩放和移

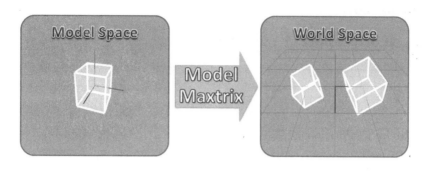

图 3 - 1 从 Model Space 到 World Space

动了位置,这是通过 2 个不同的 Model Matrix 实现的。注意,在 OpenGL 中,Model Space 和 World Space 都使用了右手系。

CSharpGL 中已经集成了对 Model Matrix 的计算。只需提供模型的方位信息,即可获得需要的 Model Matrix。例如,可以通过下述代码得到图 3-1 中较小的 Cube 所需的 Model Matrix:

```
// identity matrix at first. modelMatrix now is:
// 1 0 0 0
// 0 1 0 0
// 0 0 1 0
// 0 0 0 1
mat4 modelMatrix = mat4.identity();
// rotate model.
float degree = 30.0f;
vec3 axis = new vec3(1, 1, 1);
modelMatrix = glm.rotate(modelMatrix, degree, axis);
// scale model.
vec3 scale = new vec3(0.5f, 0.5f, 0.5f);
modelMatrix = glm.scale(modelMatrix, scale);
// translate model and we get final result.
vec3 worldPosition = new vec3(0.5f, 0, 0);
// final model matrix.
modelMatrix = glm.translate(modelMatrix, worldPosition);
```

在场景树结构中,子节点的方位继承了父节点,即子节点除了自身的旋转缩放平移外,还要按照父节点的方位调整自己。也就是说,节点的方位属性都是相对于其父节点的。当父节点平移时,子节点会随之平移;当父节点旋转或缩放时,子节点也会以父节点的中心为中心而旋转或缩放。为实现这样的效果,对模型方位的操作必须按照 Rotate→Scale→Translate 的顺序(或者 Scale→Rotate→Translate 的顺序)进行。对于子节点,必须先按照自身的 Model Matrix 变换,然后再依次按照父节点的

Model Matrix 变换。

3.2.2 World Space→View Space

View Space 也被称为 Camera Space 或 Eye Space，因为 View Space 的原点和坐标系是根据 Camera 指定的。

Camera 有三个属性：Position 描述其位置，Target 描述其朝向，Up 描述其头顶方向。Camera 的 Position 是 World Space 里的一个位置（Position.x，Position.y，Position.z），也是 View Space 里的原点；Target 也是 World Space 里的一个位置，从 Target 到 Position 的方向是 View Space 的 Z 轴正方向；Up 是 1 个向量，包含了 View Space 里的 Y 轴正方向。注意，Camera 的这三个属性都是在 World Space 里定义的。

在 World Space 中，各个模型都摆放好了位置和角度，之后就要从某个位置用 Camera 去看这个场景；也就是说，要把模型在 World Space 里的位置，经过 View Matrix 的变换，变成在 View Space 里的位置。Camera 的参数（Position，Target，Up）决定了 View Matrix。

根据 Camera 的 Position、Target 和 Up 来求 View Matrix 的过程就是著名的 lookAt(..)函数，代码如下：

```
// get view matrix.
public static mat4 lookAt(vec3 position, vec3 target, vec3 upVector)
{
    // camera's back in world space.
    vec3 back = (position - target).normalize();
    // camera's right in world space.
    vec3 right = upVector.cross(back).normalize();
    // camera's up in world space.
    vec3 up = back.cross(right);

    mat4 viewMat = mat4.identity();
    // Rotation in world space.
    viewMat.col0.x = right.x; viewMat.col1.x = right.y; viewMat.col2.x = right.z;
    viewMat.col0.y = up.x;    viewMat.col1.y = up.y;    viewMat.col2.y = up.z;
    viewMat.col0.z = back.x;  viewMat.col1.z = back.y;  viewMat.col2.z = back.z;

    // Translation in world space.
    viewMat.col3.x = - position.dot(right);
    viewMat.col3.y = - position.dot(up);
    viewMat.col3.z = - position.dot(back);

    return viewMat;
}
```

此函数中的 right/up/back 指的就是 Camera 的右侧、上方、后面。right/up/back 是三个互相垂直的单位向量(构成一个右手系),且是在 World Space 中描述的,如图 3-2 所示。

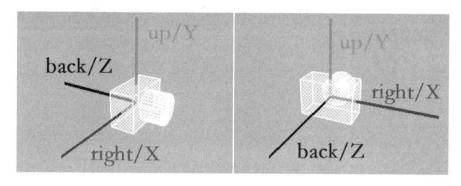

图 3-2 Camera 与 View Space 坐标系

函数得到的结果 viewMatrix 可以用式(3-9)描述。其中 right/up/back 分别描述了 Camera 坐标系的 X/Y/Z 轴,且在 viewMatrix 里也依次位于第 0/1/2 行。

$$\begin{bmatrix} right.x & right.y & right.z \\ up.x & up.y & up.z \\ back.x & back.y & back.z \\ 0 & 0 & 0 \end{bmatrix} - \begin{bmatrix} right.x & right.y & right.z \\ up.x & up.y & up.z \\ back.x & back.y & back.z \end{bmatrix} \begin{bmatrix} position.x \\ position.y \\ position.z \end{bmatrix}$$

(3-9)

从式(3-9)可以看出,viewMatrix 左上方的内容,本质上是直接将 Camera 的属性放入其中。

从 World Space 到 View Space 的过程与从 Model Space 到 World Space 的过程类似,仅从矩阵运算角度看,完全可以合并为一次变换。但场景是固定的,而 Camera 可以改变,甚至可以有多个,因此,分别建立 World Space 和 View Space 是有必要的。

View Space 使用的也是右手系。

3.2.3 View Space→Clip Space

Camera 摆好之后,就可以把三维世界的模型投影到 Camera 的 XY 轴所在的平面上(具体地说,是与其平行的一个平面上)。经过透视投影或正交投影之后的坐标就是在 Clip Space 里的坐标。

1. 透视投影

透视投影的效果就是近大远小,符合人眼的观察经验,如图 3-3 所示。

透视矩阵的作用就是设定图 3-4 所示的一个棱台范围,据此棱台变换 Camera

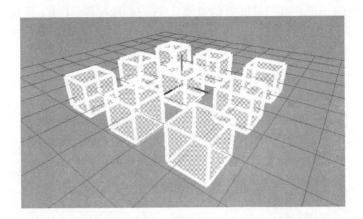

图 3-3　透视投影的效果

Space 里的顶点位置。想象这个棱台是一块海绵，里面摆放着各种物体，把这个棱台的 Far 面缓缓挤压到与 Near 面相同的大小。在这一过程中，远处的顶点比变换之前更加靠近彼此，越远的物体被挤压的程度越大，越远的物体，顶点就靠近得越多，这就是透视投影实现近大远小的原理。

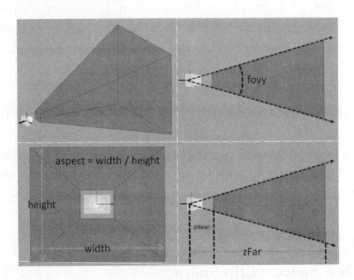

图 3-4　透视投影的棱台范围

只有位于此棱台内部的顶点才可能最终显示到画布上。在此棱台外部的顶点会被后续步骤裁切掉。

根据描述棱台的参数计算透视投影矩阵的函数就是著名的 perspective(..) 函数，其代码如下：

```
// Creates a perspective transformation matrix.
public static mat4 perspective(float fovy, float aspect, float zNear, float zFar)
{
    float tangent = (float)Math.Tan(fovy / 2.0f);
    float height = zNear * tangent;
    float width = height * aspect;
    float left = - width, right = width, bottom = - height, top = height, near =
            zNear, far = zFar;

    mat4 result = frustum(left, right, bottom, top, near, far);

    return result;
}

// Creates a frustrum projection matrix.
public static mat4 frustum(float left, float right, float bottom, float top, float near,
                    float far)
{
    mat4 result = mat4.identity();

    result[0, 0] = (2.0f * near) / (right - left);
    result[1, 1] = (2.0f * near) / (top - bottom);
    result[2, 0] = (right + left) / (right - left);
    result[2, 1] = (top + bottom) / (top - bottom);
    result[2, 2] = -(far + near) / (far - near);
    result[2, 3] = -1.0f;
    result[3, 2] = -(2.0f * far * near) / (far - near);
    result[3, 3] = 0.0f;

    return result;
}
```

2. 正交投影

正交投影没有近大远小的效果,无论远近都是同样的大小,如图 3-5 所示。

正交矩阵的作用是设置一个长方体范围,据此变换 Camera Space 里的顶点位置,如图 3-6 所示。

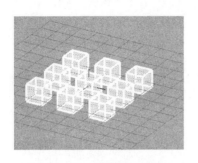

图 3-5　正交投影的效果

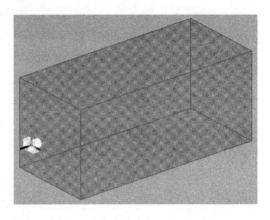

图 3-6　正交投影的立方体范围

类似透视投影，只有位于此长方体内部的顶点才可能最终显示到窗口上。此长方体外部的顶点会被后续的步骤裁切掉。

根据正交投影的参数计算正交投影矩阵的函数就是著名的 ortho(..) 函数，其代码如下：

```
// Creates a matrix for an orthographic parallel viewing volume.
public static mat4 ortho(float left, float right, float bottom, float top, float zNear,
                float zFar)
{
    var result = mat4.identity();
    // result[colume, row] = ...
    result[0, 0] = (2f) / (right - left);
    result[1, 1] = (2f) / (top - bottom);
    result[2, 2] = -(2f) / (zFar - zNear);
    result[3, 0] = -(right + left) / (right - left);
    result[3, 1] = -(top + bottom) / (top - bottom);
    result[3, 2] = -(zFar + zNear) / (zFar - zNear);
    return result;
}
```

3. 性　质

无论是透视投影还是正交投影，都有以下性质：

◆ 在 Clip Space 里的顶点位置 (x, y, z, w)，就是在 Vertex Shader 里赋值给 gl_Position 的值；

◆ 在 Clip Space 里的顶点位置 (x, y, z, w)，"x, y, z 的绝对值都小于等于 $|w|$"等价于"此顶点在可见范围之内"。

3.2.4 Clip Space → Normalized Device Space

从 Clip Space 到 Normalized Device Space(NDS)很简单,只需将(x,y,z,w)全部除以 w 即可,如式(3-10)所示。这被称为视角除法(perspective division)。

$$\begin{bmatrix} x_{nds} \\ y_{nds} \\ z_{nds} \\ w_{nds} \end{bmatrix} = \begin{bmatrix} x_{clip}/w_{clip} \\ y_{clip}/w_{clip} \\ z_{clip}/w_{clip} \\ w_{clip}/w_{clip} \end{bmatrix} \quad (3-10)$$

此时,新坐标的 w_{nds} 分量全部为1,不再需要记录。

如3.2.3小节所述,在可见范围内的(x,y,z,w),x、y 和 z 的绝对值都小于等于$|w|$。因此,经过视角除法后,所有在可见范围内的顶点位置都介于$[-1,-1,-1]$和$[1,1,1]$之间。视角除法这一过程由 OpenGL Pipeline 自动完成,开发者无需也不能参与。

3.2.5 Normalized Device Space → Window Space

最后一步,就是把$[-1,-1,-1]$和$[1,1,1]$之间的顶点位置转换到二维的屏幕窗口上,即从 Normalized Device Space 变换到 Window Space(或 Screen Space)。

假设用于 OpenGL 渲染的画布的宽、高分别为 Width、Height。glViewport(int x, int y, int width, int height);用于指定将渲染结果铺到画布的某一块区域上。一般情况下会用 glViewport(0, 0, Width, Height)来告诉 OpenGL:把结果渲染到整个画布上。如果画布尺寸发生改变,就需要再次调用 glViewport(0, 0, Width, Height)。还有一个不常见的 glDepthRange(float near, float far);用于指定在 Window Space 上的 Z 轴坐标范围(默认范围是 glDepthRange(0, 1))。

当使用 glViewport(x, y, Width, Height)的设定时,Window Space 的原点就在画布的(x,y)处。注意,Window Space 也是有第三个坐标轴 Z 的,且其方向是从计算机窗口指向里面,而其 X 轴指向右方,Y 轴指向上方。也就是说,Window Space 的 X、Y、Z 轴构成了一个左手系,这与之前的 Space 不同。

顶点在 Window Space 里的位置是按式(3-11)计算的。这一步也由 OpenGL 自动完成,开发者无需也无法参与。

$$\begin{bmatrix} x_{ws} \\ y_{ws} \\ z_{ws} \end{bmatrix} = \begin{bmatrix} \dfrac{width}{2} x_{nds} + \left(x + \dfrac{width}{2}\right) \\ \dfrac{height}{2} y_{nds} + \left(y + \dfrac{height}{2}\right) \\ \dfrac{far - near}{2} z_{nds} + \dfrac{far + near}{2} \end{bmatrix} \quad (3-11)$$

式（3-11）很简单，根据 NDS 和 Window Space 里坐标的线性关系可知式（2-12）所示关系：

$$\begin{cases} -1 \to x \\ 1 \to x+\text{width} \end{cases} \begin{cases} -1 \to y \\ 1 \to y+\text{height} \end{cases} \begin{cases} -1 \to \text{near} \\ 1 \to \text{far} \end{cases} \quad (3-12)$$

在 WinForm 系统中，控件本身的(0,0)位置是控件的左上角。即在 mouse_down 等事件中的(e.X, e.Y)是以左上角为原点，向右为 X 轴正方向，向下为 Y 轴正方向的。所以根据 WinForm 里的 (e.X, e.Y) 计算 Window Space 里的坐标时首先要用(e.X, Height-1-e.Y)转换。

如果用软件截图工具对 WinForm 窗口截图，得到窗口图片的 Width、Height 很可能不等于用 glViewport() 设定的 Width、Height。截图得到的图片尺寸受显示器分辨率的影响，不同的显示器得到的结果不尽相同。而用 glViewport() 设定的尺寸无论在哪个显示器上都应是一致的。

3.3 Pipeline 中的空间

在图 1-1 中的 Pipeline 由三个阶段组成，它们使用的空间各不相同。

1. Prepare Data

模型数据都是在 Model Space 设定的。为了便于旋转和缩放，一般将模型的中心位置设定为原点(0,0,0)。

2. Vertex Shader

Vertex Shader 能够接收来自 Model Space 的顶点位置，然后一步步变换为 Clip Space 的顶点位置，这是在渲染一般的三维模型时常用的方式。

Vertex Shader 作为程序，当然也可以直接接受 World Space、View Space 或 Clip Space 中任意一种空间的顶点位置。在渲染二维的 UI 控件时，直接提供 Clip Space 的顶点位置就是一种很好的选择。

3. Fragment Shader

Fragment Shader 接收的是 Window Space 的顶点，在 Fragment Shader 中的内置变量 vec4 gl_FragCoord; 可以获取此片段的位置，用于着色计算。gl_FragCoord.x 表示此片段在画布 X 轴上的坐标，gl_FragCoord.y 表示此片段在画布 Y 轴上的坐标，gl_FragCoord.z 表示此片段的深度值。如果画布的宽度为 width，高度为 height，那么用 vec3(gl_FragCoord.x / width, gl_FragCoord.y / height, 0) 表示片段颜色值的渲染效果如图 3-7 所示。

如图 3-7 所示，画布左下角的颜色为黑色(0,0,0)，表明此处的 gl_FragCoord.xy 为(0,0)；画布右上角的颜色为黄色(1,1,0)，表明此处的 gl_FragCoord.xy 为(width, height)。这也印证了 gl_FragCoord.xy 表示的是片段在 Window Space 中

图 3-7 对 gl_FragCoord 的实验

的坐标。读者可在本书网络资料上的 c03d06_FragCoord 项目中找到此示例的完整代码。

3.4 实例化渲染

如果场景中有很多相同或相似的物体,那么为它们分别建立模型、Shader 和 Node 等数据就显得冗余和浪费。OpenGL 提供了一个被称为实例化渲染(Instanced Rendering)的方法来避免这样的浪费。本节以一个简单的示例说明如何使用这一方法。本示例将 100 个四边形均匀地渲染到画布的各个位置上。

对于使用 glDrawArrays(..)渲染命令的模型,可以使用其对应的 Instanced 版本,即:

```
glDrawArraysInstanced(uint mode, int first, int count, int primCount);
```

当 primCount 为 100 时,意味着会将指定的模型渲染 100 次。
本示例使用的 Vertex Shader 和 Fragment Shader 如下:

```
// vertex shader.
#version 330 core

in vec2 inPosition;
in vec3 inColor;

out vec3 passColor;

uniform vec2 offsets[100];

void main()
{
```

```glsl
    vec2 offset = offsets[gl_InstanceID];
    gl_Position = vec4(inPosition + offset, 0.0, 1.0);
    passColor = inColor;
}

// ----------------------------------------------------
// fragment shader.
#version 330 core
out vec4 outColor;

in vec3 passColor;

void main()
{
    outColor = vec4(passColor, 1.0);
}
```

在 Vertex Shader 中的系统内置变量 int gl_InstanceID; 记录了此次执行的 Vertex Shader 在 Instanced 版本的渲染命令中的位置,即当前是第几遍渲染模型。如果使用的是非 Instanced 版本的渲染命令,gl_InstanceID 的值始终为 0。

在每一次渲染模型时,Vertex Shader 会根据 gl_InstanceID 的不同而取出不同的偏移量,从而将模型渲染到不同的位置上。本例中直接指定了模型的各个顶点在 Clip Space 中的偏移量。在实际应用中,可以指定顶点在任何适宜的 Space 中的偏移量。注意,偏移量是一个 uniform 的数组类型,应该在客户端设置其内容,代码如下:

```csharp
protected override void DoInitialize()
{
    base.DoInitialize();

    ShaderProgram program = this.RenderUnit.Methods[0].Program;
    var translations = new vec2[100];
    int index = 0;
    float offset = 0.1f;
    for (int y = -10; y < 10; y += 2)
    {
        for (int x = -10; x < 10; x += 2)
        {
            var item = new vec2((float)x / 10.0f + offset, (float)y / 10.0f + offset);
            translations[index++] = item;
        }
    }
    program.SetUniform("offsets", translations);
```

将四边形渲染 100 次到画布的各个位置上的效果如图 3-8 所示。

图 3-8　实例化渲染(Instanced Rendering)效果图

综上所述,在使用 Instanced 版本的渲染命令后,可以根据 Shader 中的内置变量 gl_InstanceID 来对不同批次的渲染执行不同的操作,从而渲染出相同或相似的物体。

3.5　总　结

● 矩阵是什么？

从学习 OpenGL 的角度讲,矩阵是多元一次方程组的简便写法。

● 三维模型在 OpenGL 中依次经历哪些空间？

Model Space → World Space → View Space → Clip Space → NDS → Window Space。

3.6　问　题

带着问题实践是学习 OpenGL 的最快方式。这里给读者提出几个问题,作为抛砖引玉之用。

1. 仿照本章对二维矩阵的推导过程,动手编写三维矩阵的乘法运算法则的推导过程。

2. 仿照本章对二维矩阵的推导过程,动手编写四维矩阵的乘法运算法则的推导过程。

第 4 章

几何着色器

学完本章之后,读者将能:
- 理解加入几何着色器(Geometry Shader)之后的 Pipeline;
- 用几何着色器渲染模型的法线。

4.1 介 绍

如果在创建 ShaderProgram 对象时添加了 Geometry Shader 的源代码,就会激活 OpenGL Pipeline 中执行 Geometry Shader 的步骤,此时的 Pipeline 如图 4-1 所示。

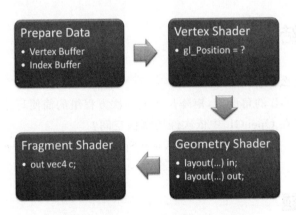

图 4-1 激活 Geometry Shader 的 Pipeline

如图 4-1 所示,Geometry Shader 接收了 Vertex Shader 的运算结果,并代替 Vertex Shader 向后续的 Fragment Shader 提供顶点信息。Geometry Shader 能够根据 Vertex Shader 的运算结果产生新的顶点和图元(PRIMITIVE),能够修改图元类型。这一功能使得 Geometry Shader 可以产生特殊的渲染效果。这里的图元是与渲染命令中的渲染模式参数匹配的基本图形类型。例如,GL_POINTS 模式下渲染的图元都是点,GL_LINES、GL_LINE_STRIP 和 GL_LINE_LOOP 模式下渲染的图元都是线段,GL_TRIANGLES、GL_TRIANGL_STRIP 和 GL_TRIANGLE_FAN 模式下渲染的图元都是三角形,GL_QUADS 和 GL_QUAD_STRIP 模式下渲染的图元都是四边形等。

激活 Geometry Shader 后,Pipeline 运行的伪代码如下:

```
// Prepare data.
vec3[] positions = ... // VertexBuffer
uint[] indexes = ... // IndexBuffer

// set uniform variables in shader.
ShaderProgram.mvpMatrix = ...
ShaderProgram.color = ...

// execute vertex shader for each vertex.
vec4[] gl_Positions = new vec4[positions.Length];
for (int i = 0; i < positions.Length; i++) // for each element in VertexBuffer:
{
    var vs = new VertexShader();
    vs.inPosition = positions[i];
    vs.main(); // execute the vertex shader we provided.
    gl_Positions[i] = vs.gl_Position;
}

// execute geometry shader.
var geometryVertexes = new List<vec3>();
var geometryIndexes = new List<uint>();
for (int i = 0; i < indexes.Length; i += GeometryShader.LayoutIn.VertexCount)
{
    var gs = new GeometryShader();
    gs.gl_in = FindPrimitive(i, indexes, gl_Positions);
    gs.main(); // execute the geometry shader we provided.
    geometryVertexes.AddRange(gs.EmitedVertexes);
    geometryIndexes.AddRange(gs.EmitedPrimitives);
}

// linear interpolation.
Fragment[] fragments = LinearInterpolation(GeometryShader.LayoutOut.DrawMode, geom-
                       etryIndexes, geometryVertexes);
Framebuffer fbo = ...

// execute fragment shader.
for (int i = 0; i < fragments.Length; i++)
```

```
{
    var fs = new FragmentShader();
    fs.main(); // execute the fragment shader we provided.
    fbo.colorAttachment.pixels[fs.FragCoord] = fs.outColor;
}

UseFramebuffer(fbo);
```

每次执行 Geometry Shader,接收的都是一个完整的图元。例如,当图元是点时,GeometryShader 会接收 1 个顶点的信息;当图元是三角形时,Geometry Shader 会接收 3 个顶点的信息。注意,这些接收的顶点的寿命到此为止,将不再传递到后续阶段。

每次执行 Geometry Shader,都可以创建数目不同的顶点和图元。这一特性十分灵活,但也破坏了 Pipeline 对数据的并行计算模式,因此在一定程度上降低了程序运行的效率。

Vertex Shader 对每个顶点、Geometry Shader 对每个图元、Fragment Shader 对每个片段。只有当需要对图元进行某种操作时,才应考虑使用 Geometry Shader。

4.2 示例:渲染法线

渲染模型的法线是 Geometry Shader 的一个典型应用。在使用高级光照算法时,模型的法线是必须使用的参数。而此例能够直观地渲染出模型的法线,便于开发者观察模型的法线数据是否正常。

本节只介绍此例的关键点,其他部分可以参考其他章节中关于 Cube 的示例,或者直接在本书网络资料上查找 Demos\Normal 项目。

4.2.1 Shaders

Vertex Shader 代码如下:

```
// vertex shader.
#version 150 core

in vec3 inPosition;
in vec3 inNormal;

out vec3 passNormal;
```

```
voidmain()
{
    gl_Position = vec4(inPosition, 1.0f);
    passNormal = inNormal;
}
```

首先,inPosition 和 inNormal 分别指代了描述顶点位置和法线的 VertexBuffer, VertexBuffer 可以简单地被视为 GPU 端的一个数组,这与无 Geometry Shader 的用法相同。然后,定义了变量 passNormal ,用于向 Geometry Shader 传递数据。

最后,在 main 函数中,如往常一样向内置变量 gl_Position 和自定义的变量 passNormal 赋值。注意,此时并未进行坐标变换,因为这将在下一步的 Geometry Shader 中实现。此时的 gl_Position 持有的不再是 Clip Space 里的坐标。Geometry Shader 代码如下:

```
// geometry shader.
#version 150 core

layout (triangles) in; // input type.
layout (line_strip, max_vertices = 2) out; // output type and vertex count.

uniform float normalLength = 0.2;
uniform vec3 vertexColor = vec3(1, 1, 1);
uniform vec3 pointerColor = vec3(0.5, 0.5, 0.5);
uniform mat4 mvpMat;

in vec3 passNormal;

out vec3 passColor; // pass color info to fragment shader.

void main()
{
    for (int i = 0; i < gl_in.length(); i++) // for each vertex in the primitive.
    {
        vec4 position = gl_in[i].gl_Position;
        passColor = vertexColor;
        gl_Position = mvpMat * position;
        EmitVertex(); // generate vertex0 of normal line.

        passColor = pointerColor;
```

```
            vec4 target = vec4(position.xyz + normalize(passNormal[i]) * normalLength,
            1.0f);
            gl_Position = mvpMat * target;
            EmitVertex(); // generate vertex1 of normal line.

            EndPrimitive();
        }
    }
```

首先,Geometry Shader 用 layout 标识符告诉 OpenGL,要接收的图元类型是三角形,要输出的则是用 GL_LINE_STRIP 模式渲染的线段。且对每个要输出的图元,最多输出 2 个顶点(max_vertices = 2),所以实际上输出的只会是一个线段。

然后,为承接 Vertex Shader 的自定义变量,这里也定义了 passNormal[]。注意,Vertex Shader 对每个顶点执行一次,Geometry Shader 对每个图元执行一次,一个图元可能有多个顶点,因此这里的变量 passNormal[] 是数组形式的。

Geometry Shader 要向后续步骤(即 Fragment Shader)提供自定义变量 passColor,这实际上是代替了 Vertex Shader 向后续步骤提供自定义变量的功能。从这一点看,可以将 Vertex Shader 和 Geometry Shader 视为一个整体,而其中的 Geometry Shader 是一个可选项。

在 main 函数中,遍历内置变量 gl_in[] 里的顶点。本例中接收的图元是三角形,因此 gl_in.length() 的值实际上为 3。本例中,对每个顶点,根据其位置和法线方向,分别产生 2 个顶点,并构成一条线段。

Fragment Shader 代码如下:

```
// fragment shader.
#version 150 core

in vec3 passColor; // color info passed from geometry shader.
out vec4 outColor;

void main()
{
    outColor = vec4(passColor, 1);
}
```

首先,passColor 承接了来自 Geometry Shader 的自定义变量 passColor。注意,Geometry Shader 对每个图元执行一次,Fragment Shader 对每个片段执行一次,每个片段只有一组顶点属性,因此这里的变量 passColor 不是数组。

在 main 函数中,将指定的颜色赋值给输出变量即可。

4.2.2 其他

本例中,Vertex Shader 需要顶点的位置和法线 2 种属性。本书使用的示例模型 annulus 的法线如图 4-2(左)所示。

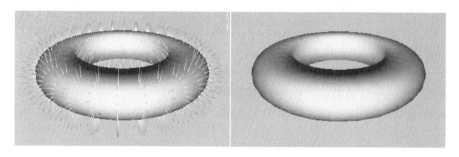

图 4-2 (左)显示顶点的法线 (右)不显示顶点的法线

注意,模型 annulus 本身只有图 4-2(右)的数据,不包含图 4-2(左)的法线线段,这些法线线段是通过 Geometry Shader 生成的。

另外,Geometry Shader 接收了模型的原始顶点,然后产生了新顶点,额外带来的效果:在 Geometry Shader 执行完毕后,原始顶点就不存在了。因此,图 4-2(左)除了用 Geometry Shader 渲染法线线段外,还用另一个 RenderMethod 渲染了模型本身。

4.3 总 结

本章向 Pipeline 引入了 Geometry Shader 这一步骤。Geometry Shader 可以视为对 Vertex Shader 的补充。当开发者需要一个图元一个图元地处理模型时,就应考虑使用 Geometry Shader。

4.4 问 题

带着问题实践是学习 OpenGL 的最快方式。这里给读者提出问题,作为抛砖引玉之用。

1. 修改 Geometry Shader,使之产生的法线线段变为彩色的箭头。
2. 使用 Geometry Shader 后,如果不渲染模型原有的样子,会有哪些应用?

第 5 章

光 照

学完本章之后,读者将能:
- 了解 OpenGL 使用的光照(Lighting)模型;
- 用 Shader 分别实现三种光源的光照;
- 用 Shader 实现同时开启多个光源的光照。

5.1 Blinn-Phong 光照模型

本章不涉及透明物体的问题。

用计算机精准再现现实中的光照情形是极其复杂的,因此 OpenGL 用一种相对简单的方式来近似地表现光照效果,那就是 Blinn-Phong 光照模型。

根据自然规律,场景中的物体,除非自身能够发光,剩下的物体都是通过反射光源光线来显示自身颜色的。

5.2 光 源

本章介绍三种类型的光源:
- 点光源(Point Light):点光源模拟的是电灯泡这一类型的光源,将电灯泡抽象为仅从一点发光,且发光范围360°无死角;
- 平行光(Directional Light):平行光模拟的是太阳光。由于太阳距离地球太远,其光线可以被视为平行地照射到地球上;
- 聚光灯(Spot Light):聚光灯模拟的是舞台上的聚光灯,可以看作是限定了照射角度的点光源。

5.3 反射光

Blinn-Phong 将物体反射的光分为三部分:环境光(Ambient)、漫反射光(Diffuse)和镜面反射光(Specular)。这三部分相加,就得到了物体最终的颜色。在 Fragment Shader 中计算光照效果的伪代码如下:

```glsl
//Blinn-Phong-WorldSpace.frag
#version 150
// light source.
struct Light {
    vec3 position;
    vec3 diffuse;
    vec3 specular;
};
// Model's property about reflecting light.
struct Material {
    vec3 diffuse;
    vec3 specular;
};

uniform Light light; // The light source variable.
uniform Material material; // The model's material variable.
// Calculate diffuse and specular factors at specifiedposition.
void LightUp(..., out float diffuse, out float specular) {
}
// output color
out vec4 outColor;

void main() {
    float diffuse = 0;
    float specular = 0;
    LightUp(..., diffuse, specular); // Gets diffuse and specular factor.

    outColor = vec4(ambient
            + diffuse * light.diffuse * material.diffuse
            + specular * light.specular * material.specular, 1.0);
}
```

不同的物体,反射光的能力不同。物体的 Material 属性描述了物体反射光线的 Diffuse 和 Specular 分量的能力。同时,不同的光源能够投射的 Diffuse 和 Specular 的分量也不同。光源的定义 Light 包含了其投射能力。光源的投射能力,物体的反射能力,光线、物体和观察者的相对位置共同决定了观察者看到的最终颜色。

5.3.1 环境光

在场景中,光线经过多次反射,会使整个场景都有一定的亮度。OpenGL 用一个简单的常量来模拟这一情形。这种光照效果被称为环境光。实现环境光的 Fragment Shader 代码如下:

```
// fragment shader.
#version 150 core

uniform vec3 ambientColor = vec3(0.2, 0.2, 0.2);

out vec4 outColor;

void main()
{
    outColor = vec4(ambientColor, 1);
}
```

显然,单纯的环境光只能让模型各处具有相同的颜色。实际上在第 1 章渲染 Cube 模型的示例中,就使用了环境光(也只使用了环境光,导致 Cube 模型呈现为单一的颜色)。

5.3.2 漫反射光

光照射到物体表面的某点,当光线方向与物体在此点的法线重合时,物体反射出的光强度最大。当光线方向与物体在此点的法线垂直时,说明光线没有照射到此点(擦肩而过),那么此点就是黑色的。物体反射的光会均匀地朝向各个方向,这种光被称为漫反射光,如图 5-1 所示。图 5-1 表现了点光源照射到一个平面上的情形。当入射光线 L 与法线 N 夹角较小时(图 5-1(左)),物体能够反射的光强度较大;当入射光线 L 与法线 N 夹角较大时(图 5-1(右)),物体能够反射的光强度较小。

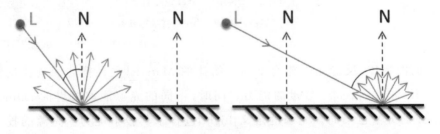

图 5-1　(左)漫反射夹角小则反射强度大(右)夹角大则反射强度小

在介绍 Geometry Shader 时使用的示例,就是使用漫反射光渲染了模型本身,这使得模型表现出很强的立体感。计算漫反射光强度的 Shader 代码如下：

```
float GetDiffuse(vec3 lightDir, vec3 normal)
{
    lightDir = normalize(lightDir);
    normal = normalize(normal);
    float diffuse = max(dot(lightDir, normal), 0);

    return diffuse;
}
```

注意,漫反射光强度的算法对于各种类型的光源都适用。

5.3.3 镜面反射光

光照射到物体表面的某点,当反射出的光线 R 与视线 v 重合时,会有特别闪亮的效果。随着反射光线与视线的偏离,这种闪亮效果会急速衰减至消失。物体反射的这种光被称为镜面反射光,如图 5-2 所示。

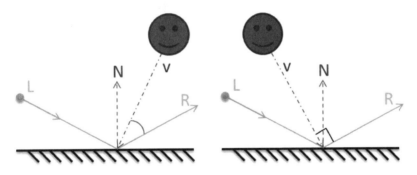

图 5-2　(左)镜面反射视线 v 与反射光线 R 呈锐角　(右)视线 v 与反射光线 R 呈直角

显然,可以通过反射光线 R 与视线 v 的夹角得到镜面反射光的强度,这是 Phong 光照模型的算法。

另一种镜面反射光的计算方式:用视线 v 与入射光线 L 的中间角 H 为准绳,根据它与法线 N 的夹角计算镜面的反射强度,如图 5-3 所示,这是 Blinn-Phong 光照模型的算法。

这里有一个特殊的性质:∠HON 是 ∠VOR 的一半。简单的证明如下:

对于图 5-3 的(a)图,有 ∠VOR = ∠NOR − ∠NOV = ∠NOL − ∠NOV = ∠HON + ∠HOL − ∠NOV = ∠HON + ∠HOV − ∠NOV = ∠HON + ∠HON + ∠NOV − ∠NOV = 2∠HON。

对于图 5-3 的(b)图,有 ∠VOR = ∠NOR + ∠NOV = ∠NOL + ∠NOV = ∠HON + ∠HOL + ∠NOV = ∠HON +

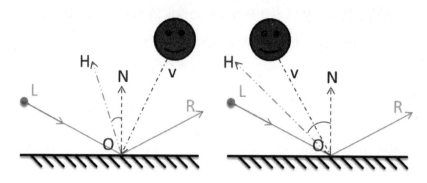

图 5-3 Blinn-Phong 的镜面反射

∠HON = 2∠HON。

因此,在 Blinn-Phong 模型中使用的∠HON 常被称为半角(halfway angle),向量 $\overrightarrow{OH}$ 则被称为半角向量。

计算镜面反射强度的 Shader 代码如下:

```
void LightUp(vec3 lightDir, vec3 normal, vec3 eyePos, vec3 vPos, float shiness, out float diffuse, out float specular) {
    // Diffuse factor
    diffuse = max(dot(lightDir, normal), 0);

    // Specular factor
    vec3 eyeDir = normalize(eyePos - vPos);
    specular = 0;
    if (blinn) { // Blinn-Phong lighting model.
        if (diffuse > 0) {
            vec3 halfwayDir = normalize(lightDir + eyeDir);
            specular = pow(max(dot(normal, halfwayDir), 0.0), shiness);
        }
    }
    else { // phong lighting model.
        if (diffuse > 0) {
            vec3 reflectDir = reflect(-lightDir, normal);
            specular = pow(max(dot(eyeDir, reflectDir), 0.0), shiness);
        }
    }
}
```

计算镜面反射光比漫反射光稍微复杂些。首先要保证漫反射光强度大于零,即光源是在模型正面,能够照射到模型上。除了光线方向和法线方向外,还需要根据观察者的位置 eyePos 来计算镜面反射强度。最后,使用 shiness 变量和指数函数

`pow()`来调节镜面反射的衰减速度。

计算光源 L 的反射光线 R，这是一个比较费时的操作。而半角向量同时包含了光源 L 和观察者 v 的方向，且计算十分简便，因此广泛用于各种光照模型。

由此代码可见，漫反射光强度和镜面反射光强度可以在同一函数中计算。这能够避免重复计算。使用上述算法计算（环境光＋漫反射光＋镜面反射光）的光照模型，被称为 Blinn-Phong 光照模型。这是 OpenGL 内部使用的光照模型。

图 5-4 展示了不同的反射光分量及其组合呈现的效果。可见，单独的 Ambient 反射光会让模型呈现为单一的颜色；单独的 Diffuse 反射光可以让模型的各个面很好地区分开；单独的 Specular 反射光会让模型在特殊的角度上呈现高亮色，其他位置则保持黑暗；三者组合所呈现的模型外观可以相当自然。

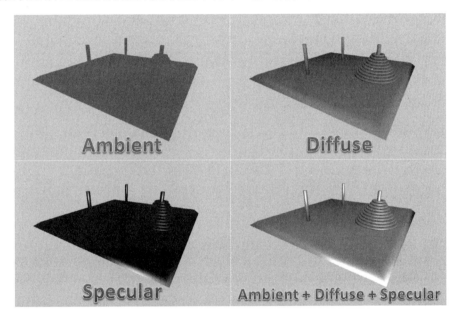

图 5-4　反射光的效果

下面分别提供三种常见光源所对应的 Blinn-Phong 算法。

5.4　Blinn-Phong 算法

5.4.1　点光源的 Blinn-Phong 算法

点光源的 Blinn-Phong 算法如下：

```
inVertex {
    vec3 position;
    vec3 normal;
}vertexIn; // vertex attributes passed from vertex shader.

// Gets diffuse and specular for point light.
void PointLightUp(Light light, out float diffuse, out float specular) {
    vec3 lvDirection = light.position - vertexIn.position;
    vec3 lightDir = normalize(lvDirection);
    vec3 normal = normalize(vertexIn.normal);
    float distance = length(lvDirection);
    float attenuation = 1.0 / (light.constant + light.linear * distance + light.quadratic * distance * distance);

    LightUp(lightDir, normal, eyePos,vertexIn.position, material.shiness, diffuse, specular);

    diffuse = diffuse * attenuation;
    specular = specular * attenuation;
}
```

一般的,点光源的光线强度会随着距离的增加而衰减,此算法考虑了衰减的程度。当衰减到一定程度时,实际上是可以忽略不计的。这一思想可以用来对光照计算进行优化。

5.4.2 平行光的 Blinn-Phong 算法

平行光的 Blinn-Phong 算法如下:

```
inVertex {
    vec3 position;
    vec3 normal;
}vertexIn; // vertex attributes passed from vertex shader.

// Gets diffuse and specular for point light.
void DirectionalLightUp(Light light, out float diffuse, out float specular) {
    vec3 lightDir = normalize(light.direction);
    vec3 normal = normalize(vertexIn.normal);
    LightUp(lightDir, normal, eyePos,vertexIn.position, material.shiness, diffuse, specular);
}
```

平行光可以看作无限远离场景的点光源,因为一般模拟的是太阳光,所以不考虑衰减问题,其光照计算十分简单。

5.4.3 聚光灯的 Blinn-Phong 算法

聚光灯的 Blinn-Phong 算法如下:

```
inVertex {
    vec3 position;
    vec3 normal;
}vertexIn; // vertex attributes passed from vertex shader.

// Gets diffuse and specular for point light.
// Note: We assume that spot light's angle ranges from 0 to 180 degrees.
void SpotLightUp(Light light, out float diffuse, out float specular) {
    vec3 lvDirection = light.position - vertexIn.position;
    vec3 lightDir = normalize(lvDirection);
    vec3 centerDir = normalize(light.direction);
    float c = dot(lightDir, centerDir);// cut off at this point.
    if (c < light.cutOff) { // current point is outside of the cut off edge.
        diffuse = 0; specular = 0;
    }
    else {
        vec3 normal = normalize(vertexIn.normal);
        float distance = length(lvDirection);
        float attenuation = 1.0 / (light.constant + light.linear * distance + light.quadratic * distance * distance);

        LightUp(lightDir, normal, eyePos,vertexIn.position, material.shiness, diffuse, specular);
        diffuse = diffuse * attenuation;
        specular = specular * attenuation;
    }
}
```

聚光灯的特点是,有一个特定的照射范围,在范围内的顶点才接收光照。

5.5 同时使用多个光源

为每个光源执行一遍渲染过程,分别计算物体在各个光源下的颜色,并将这些颜色值叠加,即为多个光源的光照效果。使用 OpenGL 的混合功能,就可以实现对颜色的叠加。

5.5.1 混　合

在 Pipeline 的 Fragment Shader 执行完毕后，产生的每个片段还要经历一系列的操作或测试。只有成功完成全部操作或测试的片段才能显示到画布上，成为像素。混合(Blend)功能就是其中一项操作。

混合是按照某种计算方法，将新片段的颜色与画布上已有的颜色混合，成为画布上的新颜色。新片段的颜色被称为源颜色(Source Color)，画布上已有的颜色被称为目标颜色(Destination Color)。这两种指代容易混淆，读者可以按下述方式记忆："源颜色"的"源"，指的是让画布上的颜色发生改变的"源头"，即 Source of Change。有了新片段，才会有改变，新片段就是改变的源头。因此，新片段的颜色被称为源颜色。

使用混合功能，必须指定混合的方式，即使用下述 OpenGL 函数：

```
glBlendFunc(uint sfactor, uint dfactor);
glBlendFuncSeparate(uint srcRGB, uint dstRGB, uint srcAlpha, uint dstAlpha);
```

一般的，在混合场景中的不同物体，使用下述参数配置混合计算方式：

```
glBlendFunc(GL_SRC_ALPHA, GL_ONE_MINUS_SRC_ALPHA);
```

如果按照从远到近的顺序渲染各个物体，就可以得到半透明的效果；但如果渲染物体的远近顺序是乱序的，那么就可能产生诡异的效果，这是混合功能的局限性。在混合多个光源的照射效果时，由于每次渲染的物体都是同一场景中同样的物体，因此在多次渲染时，同一个位置上的片段的深度值是不变的，这样就保证了混合功能的结果不会出错。

5.5.2 逐光源计算

同时启用多个光源的 Blinn-Phong 渲染算法伪代码如下：

```
// binn-phong lighting with multiple lights.
void BlinnPhongRender(Scene scene, ..)
{
    // Render ambient color first.
    RenderAmbientColor(scene.RootNode, ..);

    // Enable blend.
    glEnable(GL_BLEND); glBlendFunc(GL_ONE, GL_ONE);
    // Render scene under each light.
    foreach (var light in scene.Lights)
    {
        var arg = new RenderEventArgs(.., scene.Camera);
        RenderBlinnPhongNode(scene.RootNode, arg, light);
    }
```

```
    // disable blend
    glDisable(GL_BLEND);
}

// Render ambient color for the node and its chldren.
void RenderAmbientColor(SceneNodeBase node, BlinnPhongAmbientEventArgs arg)
{
    var blinnPhongNode = node as IBlinnPhong;
    if (blinnPhongNode != null) // if this node supports IBlinnPhong interface.
    {
        // Render this node
        blinnPhongNode.RenderAmbientColor(arg);
    }

    // Render children
    foreach (var child in node.Children)
    {
        RenderAmbientColor(child, arg);
    }
}

// Render diffuse color and specular color for the node and its children under specified
   light
void RenderBlinnPhongNode(SceneNodeBase node, RenderEventArgs arg, LightBase light)
{
    var blinnPhongNode = node as IBlinnPhong;
    if (blinnPhongNode != null) // if this node supports IBlinnPhong interface.
    {
        // Render this node
        blinnPhongNode.Render(arg, light);
    }

    // Render children
    foreach (var child in node.Children)
    {
        RenderBlinnPhongNode(child, arg, light);
    }
}
```

此算法首先计算了场景中各个节点颜色的 Ambient 分量,且在此之后才开始启用混合模式,这就保证了场景中的物体至少都被渲染一次。如果场景中某个物体没有被任何光源照射到,它至少能够显示出昏暗的底色,且不会出现异常的透明效果。

读者可以将这一步看作是对场景中的特殊内置光源进行计算,此内置光源只产生简单的 Ambient 纯色。然后,启用混合效果,且设置混合参数为 glBlendFunc (GL_ONE, GL_ONE)。这意味着,所有的光源照射到场景中而产生的颜色,都会全部相加,这样的加法与顺序无关。这就符合了真实场景的情形。

渲染 Ambient 分量时,需要遍历场景中的所有节点。每次渲染一个光源的照射效果时,也都要遍历场景中的所有节点。显然,如果场景中的光源比较多,这样的计算会严重拖累效率。此时可以用"延迟渲染"的方法来降低需要渲染的工作量。读者可自行查找相关资料,本章不做介绍。

5.6 阴 影

光照射到物体上,不仅会产生明亮的效果,还会在地上产生阴影。从 OpenGL Pipeline 的角度思考,阴影的本质是指定某些位置,取消其光照计算。换句话说,如果能够通过某种方法记录下来所有不需要计算光照的位置,就能够达到阴影效果。

为了记录阴影信息,需要用到 Framebuffer。下一章将集中介绍 Framebuffer 的概念和用法。在此基础之上才能实现渲染阴影的效果。

5.7 凹凸映射

凹凸映射(Bump Mapping)综合利用纹理、矩阵和 Shader,是一种渲染细节真实感的方法。这一技术特别适用于整体平整而又有轻微凹凸痕迹的物体表面,例如年代久远的砖墙、大树树皮和带坑的星球等。一个应用了凹凸映射的平面示例如图右图 5-5 所示。

图 5-5 (左)常规光照 (右)凹凸映射

本节在实现此示例的过程中逐步讲解其原理。读者可在本书网络资料上的 Demos\NormalMapping 项目中找到此示例的完整代码。

凹凸映射的基本思想：轻微地修改顶点的法线，从而让物体表面在光照计算中呈现出凹凸感。这是一种"欺骗"光照模型的方法，因此不需要增加模型的顶点数量，也无法真正地让模型凹凸起来，只是在远处看起来有凹凸的感觉。这是一种优化的思想，即在不必要的情形下不增加某些不会被使用的资源（例如描述这里凹凸细节的顶点）。

一般的光照计算可以在 World Space 或 View Space 里实施，但是凹凸映射需要在一个局部表面坐标空间(Surface－local Coordinate Space)中计算顶点被修改后的法线，然后将法线变换到 World Space 或 View Space。这个空间针对每个顶点分别建立，以顶点为原点，由下述直角坐标系定义：其 Z 轴为顶点的原法线向量（Normal），X 轴用一个新增的顶点属性（切线向量 Tangent）表示，Y 轴坐标可以通过 X 轴和 Z 轴向量的叉积运算得到（称为副法线 Binormal），这个空间也被称为切空间(Tangent Space)。得到新法线后，可以用式(5－1)所示的矩阵将新法线从 Tangent Space 变换到 World Space。

$$\begin{bmatrix} tangent.x & binormal.x & normal.x \\ tangent.y & binormal.y & normal.y \\ tangent.z & binormal.z & normal.z \end{bmatrix} \quad (5-1)$$

注意，这里的顶点不仅仅是在 VertexBuffer 中设置的顶点，它包含了在 Fragment Shader 处理阶段之前通过线性插值生成的所有顶点（即片段 Fragment）。因此，为了让使用的纹理贴图在任何一点都能找到合适的新法线，需要将这些新法线也存储到一个纹理贴图中，如图 5－6 所示。

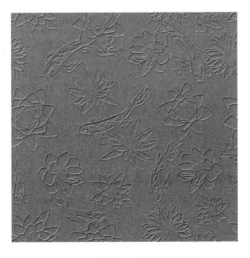

图 5－6　存储了新法线的贴图

法线贴图本质上和普通的纹理对象没有什么不同,只是其存储的RGB颜色值表示的是法线在Tangent Space中被修改后的值(RGB分别对应XYZ轴上的分量)。法线被修改的幅度通常很小,因此R和G的值一般很小,B的值一般很大,这使得法线贴图直观上看都是偏蓝色的。本示例的Vertex Shader如下:

```glsl
// vertex shader.
#version 330

in vec3 inPosition;
in vec3 inNormal;
in vec3 inTangent;
in vec2 inTexCoord;

uniform mat4 mvpMat; // projection * view * model
uniform mat4 modelMat; // model

out vec3 passWorldPos; // position in world space.
out vec3 passNormal; // normal in world space.
out vec3 passTangent; // tangent in world space.
out vec2 passTexCoord; // texture coordinate.

void main()
{
    gl_Position = mvpMat * vec4(inPosition, 1.0);
    passWorldPos = (modelMat * vec4(inPosition, 1.0)).xyz;
    passNormal = (modelMat * vec4(inNormal, 0.0)).xyz;
    passTangent = (modelMat * vec4(inTangent, 0.0)).xyz;
    passTexCoord = inTexCoord;
}
```

Vertex Shader将顶点在World Space中的位置、法线和切线属性传递到Fragment Shader,供其计算调整后的法线。本示例的Fragment Shader如下:

```glsl
// fragment shader.
#version 330

in vec2 passTexCoord;
in vec3 passNormal;
in vec3 passWorldPos;
in vec3 passTangent;

out vec4 outColor;
```

```glsl
struct DirectionalLight // light info.
{
    vec3 color;
    float ambient;
    float diffuse;
    vec3 direction;
};

uniform DirectionalLight light;
uniform sampler2D texColor; // color in regular texture.
uniform sampler2D texNormal ; // modified normal in bump texture.
uniform vec3 eyeWorldPos; // camera position.
uniform float matSpecular = 1;
uniform float specularPower = 1;

// regular lighting process.
vec4 CalcDirectionalLight(vec3 normal)
{
    vec4 ambient = vec4(light.color * light.ambient, 1.0f);
    float diffuseFactor = dot(normal, - light.direction);

    vec4 difuse = vec4(0, 0, 0, 0);
    vec4 specular = vec4(0, 0, 0, 0);

    if (diffuseFactor > 0) {
        difuse = vec4(light.color * light.diffuse * diffuseFactor, 1.0f);

        vec3 v2e = normalize(eyeWorldPos - passWorldPos);
        vec3 lightReflect = normalize(reflect(light.direction, normal));
        float specularFactor = dot(v2e, lightReflect);
        if (specularFactor > 0) {
            specularFactor = pow(specularFactor, specularPower);
            specular = vec4(light.color * matSpecular * specularFactor, 1.0f);
        }
    }

    return (ambient + difuse + specular);
}

vec3 CalcBumpedNormal()
{
```

```glsl
    vec3 normal = normalize(passNormal);
    vec3 tangent = normalize(passTangent);
    tangent = normalize(tangent - dot(tangent, normal) * normal);
    vec3 bitangent = cross(tangent, normal);
    mat3 tbnMat = mat3(tangent, bitangent, normal); // the TBN matrix.

    vec3 bumpNormal = texture(texNormal, passTexCoord).xyz;
    bumpNormal = 2.0 * bumpNormal - vec3(1.0, 1.0, 1.0); // normal in texture.

    vec3 newNormal = tbnMat * bumpNormal; // from tangent space to world space.

    return normalize(newNormal);
}

void main()
{
    vec3 normal = CalcBumpedNormal();
    vec4 lightColor = CalcDirectionalLight(normal); // regular lighting process.

    vec4 color = texture(texColor, passTexCoord.xy);
    outColor = color * lightColor;
}
```

CalcBumpedNormal()根据法线贴图中的法线数据和TBN矩阵,得到了被轻微修改后的法线(在World Space内)。用此法线参与光照计算,就可以得到有凹凸质感的效果。

5.8 噪声

OpenGL能够渲染十分精确的图形,但是有时反而需要看起来杂乱无章的图形。如果需要渲染类似云雾的图形,或者给物体添加随机性的外观,可以使用噪声技术。OpenGL Shader 内置了产生噪声的函数如下:

```glsl
// return anoise value based on input value x.
float noise1(genType x);
    vec2 noise2(genType x);
    vec3 noise3(genType x);
    vec4 noise4(genType x);
```

但是并非所有的 OpenGL 实现都支持这个函数,实现的具体方式也可能有所不同,因此最好使用一种自定义的、可移植的方法来实现噪声函数。

本节涉及的噪声函数必须具有这些性质：噪声不应具有明显的重复样式；应当平滑地变化，不会突然剧烈地改变；每次输入同样的数据，都应产生同样的结果；噪声函数不应太复杂，以免降低程序效率。

本节介绍一个噪声函数并据此模拟太阳表面的活动。读者可在本书网络资料的示例项目 Demos\SimpleNoise.Sun 找到本示例的完整代码。本示例使用的产生噪音的函数 snoise 实现如下：

```
//Description : Array and textureless GLSL 2D/3D/4D simplex noise functions
//Author : Ian McEwan, Ashima Arts.
// Maintainer : stegu
// Lastmod : 20110822 (ijm)
//License : Copyright (C) 2011 Ashima Arts. All rights reserved.
//           Distributed under the MIT License. See LICENSE file.
//           https://github.com/ashima/webgl-noise
//           https://github.com/stegu/webgl-noise
vec3 mod289(vec3 x) {
    return x - floor(x * (1.0 / 289.0)) * 289.0;
}

vec4 mod289(vec4 x) {
    return x - floor(x * (1.0 / 289.0)) * 289.0;
}

vec4 permute(vec4 x) {
    return mod289(((x * 34.0) + 1.0) * x);
}

vec4 taylorInvSqrt(vec4 r) {
    return 1.79284291400159 - 0.85373472095314 * r;
}

float snoise(vec3 v) {
    const vec2 C = vec2(1.0/6.0, 1.0/3.0) ;
    const vec4 D = vec4(0.0, 0.5, 1.0, 2.0);

    // First corner
    vec3 i = floor(v + dot(v, C.yyy) );
    vec3 x0 =   v - i + dot(i, C.xxx) ;

    // Other corners
    vec3 g = step(x0.yzx, x0.xyz);
```

```glsl
    vec3 l = 1.0 - g;
    vec3 i1 = min( g.xyz, l.zxy );
    vec3 i2 = max( g.xyz, l.zxy );
    vec3 x1 = x0 - i1 + C.xxx;
    vec3 x2 = x0 - i2 + C.yyy;
    vec3 x3 = x0 - D.yyy;

    // Permutations
    i = mod289(i);
    vec4 p = permute(permute(permute(
             i.z + vec4(0.0, i1.z, i2.z, 1.0)
           + i.y + vec4(0.0, i1.y, i2.y, 1.0))
           + i.x + vec4(0.0, i1.x, i2.x, 1.0)));

    // Gradients: 7x7 points over a square, mapped onto an octahedron.
    // The ring size 17*17 = 289 is close to a multiple of 49 (49*6 = 294)
    float n_ = 0.142857142857;
    vec3 ns = n_ * D.wyz - D.xzx;

    vec4 j = p - 49.0 * floor(p * ns.z * ns.z);

    vec4 x_ = floor(j * ns.z);
    vec4 y_ = floor(j - 7.0 * x_);

    vec4 x = x_ *ns.x + ns.yyyy;
    vec4 y = y_ *ns.x + ns.yyyy;
    vec4 h = 1.0 - abs(x) - abs(y);

    vec4 b0 = vec4( x.xy, y.xy );
    vec4 b1 = vec4( x.zw, y.zw );

    vec4 s0 = floor(b0) * 2.0 + 1.0;
    vec4 s1 = floor(b1) * 2.0 + 1.0;
    vec4 sh = - step(h, vec4(0.0));

    vec4 a0 = b0.xzyw + s0.xzyw * sh.xxyy ;
    vec4 a1 = b1.xzyw + s1.xzyw * sh.zzww ;

    vec3 p0 = vec3(a0.xy, h.x);
    vec3 p1 = vec3(a0.zw, h.y);
    vec3 p2 = vec3(a1.xy, h.z);
```

```
        vec3 p3 = vec3(a1.zw, h.w);

        //Normalise gradients
        vec4 norm = taylorInvSqrt(vec4(dot(p0, p0), dot(p1, p1), dot(p2, p2), dot(p3,
            p3)));
        p0 *= norm.x;
        p1 *= norm.y;
        p2 *= norm.z;
        p3 *= norm.w;

        // Mix final noise value
        vec4 m = max(0.6 - vec4(dot(x0, x0), dot(x1, x1), dot(x2, x2), dot(x3, x3)), 0.
            0);
        m = m * m;
        return 42.0 * dot(m * m,
            vec4( dot(p0,x0), dot(p1,x1), dot(p2,x2), dot(p3, x3)));
}
```

读者可以使用任何简单或复杂的噪声函数实现，不必纠结于本书提供的示例。由于噪声函数对于相同的输入总是给出相同的输出，而本示例要演示太阳表面的不断变化，所以必须不断地修改噪声函数的输入值。这一输入值可以用一个简单的uniform 变量表示。本示例的 Fragment Shader 代码如下：

```
// fragment shader.
#version 150 core

uniform float time; // updates each frame.
uniform sampler1D sunColor; // mapping from noise output to color.

out vec4 outColor;

void main()
{
    float x = snoise(vec3(0.0, 0.0, time));

    // A hot colormap - cheesy but effective.
    outColor = texture(sunColor, x);
}
```

Fragment Shader 代码中的 uniform 变量 time 要在客户端中的每一帧更新。这样才能让 Shader 每次渲染出不同的噪声效果。如图 5-7 所示。

本示例只需用一个普通的球体模型。在其表面赋予由噪声提供的颜色，即可表现出太阳表面的耀斑和黑子现象。

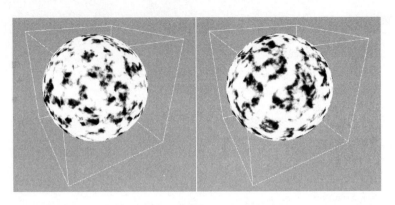

图 5-7　用噪声表现太阳表面

5.9　总　结

本章介绍了 OpenGL 内部使用的光照模型 Blinn-Phong，然后用 Shader 实现了 Blinn-Phong 算法，通过混合功能实现了对多光源的支持，简要介绍了综合运用纹理、矩阵和 Shader 实现凹凸映射的方法，最后介绍了使用噪声创建杂乱效果的一般方法。

5.10　问　题

带着问题实践是学习 OpenGL 的最快方式。这里给读者提出几个问题，作为抛砖引玉之用。

1. 聚光灯实际上是照射范围为圆锥体的光源。请读者设计实现照射范围为四棱锥的光源。

2. 请读者尝试在多光源算法中取消 Ambient 分量计算，或者将 Ambient 分量计算放到最后，观察渲染效果有何区别？

第6章

帧缓存

学完本章之后,读者将能:
- 了解帧缓存(Framebuffer)的概念和构造;
- 使用帧缓存实现 Render to Texture。

存储顶点属性的 VertexBuffer 对象和存储索引的 IndexBuffer 对象都是缓存对象,它们都可以被视为位于显卡内存中的一个数组。但 Framebuffer 不是一个缓存对象,它是若干个特定缓存对象的容器,它包含若干个缓存对象。在首次介绍 Pipeline 的伪代码中,已经出现了 Framebuffer 对象:

```
Framebuffer fbo = ...
//...
// execute fragment shader.
for (int i = 0; i < fragments.Length; i++)
{
    var fs = new FragmentShader();
    fs.main(); // execute the fragment shader we provided.
    // assign fragment shader's output to framebuffer.
    fbo.colorAttachment.pixels[fs.FragCoord] = fs.outColor;
}

UseFramebuffer(fbo);
```

这里的 Framebuffer 对象接收了来自 Fragment Shader 的输出结果,并保存到自己的颜色附件 colorAttachment(即一个记录颜色信息的缓存对象)上。之后 OpenGL 将此附件中的颜色显示到画布(或者说 WinForm 的窗口的某个控件)上,就完成了最终的显示。

每个 OpenGL 应用程序执行时,都会自动创建一个默认的 Framebuffer 对象。虽然开发者可以自行创建新的 Framebuffer 对象,但是只有这个默认的 Framebuffer 中的内容才能被渲染到窗口上。

开发者自定义的 Framebuffer 主要用于在多遍渲染算法中缓存一些中间数据。一个 Framebuffer 对象至少应包含一个记录颜色的 Color Buffer 和一个记录深度的 Depth Buffer。

6.1 名词术语

为了叙述方便,首先定义一些术语。本章在涉及下述术语时将直接使用其英文名称。这样能够避免某些中文术语可能引起的歧义。

◆ Image

记录像素颜色的二维数组(Pixel[,]),有特定的格式。

◆ Layered Image

大小和格式相同的一组 Image。

◆ Texture

包含若干 Image 的 OpenGL 对象。这些 Image 的格式相同,但大小未必相同(例如不同的 mipmap level 的 Image 大小是不同的)。Texture 可以以多种方法被 Shader 读取。

◆ Renderbuffer

包含 1 个 Image 的 OpenGL 对象。不能被 Shader 读取,只能被创建,然后放到 FBO 里。可以认为它是一个简化的 Texture 对象,其作用类似于一个占位符。

◆ Attach

把一个对象附着(Attach)到另一个对象上。附着不同于绑定(Bind),对象被绑定到上下文(Context),之后的相关操作都作用于此对象;对象被附着到另一个对象,之后的操作与此对象没有必然的关系。

◆ AttachmentPoint

Framebuffer 对象里可以让 Image 或 Layered Image 附着的位置,只有符合规定的图像格式才能被附着。

◆ Framebuffer-attachableImage

格式符合规定,可以被附着到 framebuffer 对象的 Image。

◆ Framebuffer-attachableLayered Image

格式符合规定,可以被附着到 framebuffer 对象的 Layered Image。

6.2 附着点

FBO 有若干 Image 的附着点(Attachment Point),一个附着点可以附着一个附件。

◆ GL_COLOR_ATTACHMENT*i*

这些附着点的数量依不同的 OpenGL 实现而不同。可以用 GL_MAX_COLOR_ATTACHMENTS 查询一个 OpenGL 实现支持的颜色附着点的数量。最少有 8 个附着点,因此可以放心使用附着点 GL_COLOR_ATTACHMENT0 到 GL_COLOR_

ATTACHMENT7。这些附着点只允许附着可渲染色彩的 Image。所有压缩的 Image 格式都不是可渲染色彩的,所以都不能附着到这些附着点上。

◆ GL_DEPTH_ATTACHMENT

这个附着点只允许附着 depth 格式的 Image。附着的 Image 就成为此 FBO 的 Depth Buffer。注意,即使不打算从深度附着点上读取什么信息,也应该给深度附着点设定一个 Image。此时用一个 Renderbuffer 作为 Image 就成为了一个方便的选项。

◆ GL_STENCIL_ATTACHMENT

这个附着点只允许附着 stencil 格式的 Image。附着的 Image 就成了此 FBO 的 Stencil Buffer。

◆ GL_DEPTH_STENCIL_ATTACHMENT

这是"depth+stencil"的方式,即附着的 Image 既是 Depth Buffer 又是 Stencil Buffer。注意:如果使用 GL_DEPTH_STENCIL_ATTACHMENT,应当使用一个以 packed depth-stencil 为内部格式(internal format)的 Texture 或 Renderbuffer 充当附件。

6.3 将 Texture 附着到 Framebuffer

OpenGL 支持多种类型的 Texture。一个 Texture 对象里可能包含多个 Image。而将 Texture 附着到 Framebuffer 时,实际上是将 Texture 里的某个 Image 附着到 Framebuffer。

能够附着到 Framebuffer 的 Texture 类型如下所述。

6.3.1 附着 1D Texture

创建一维纹理(1D Texture)对象时,只需指定其长度,不需指定其宽度。可以视其为宽度为 1 的二维纹理对象。一维纹理同二维纹理一样,也可以由多个 level 构成,如图 6-1 所示。

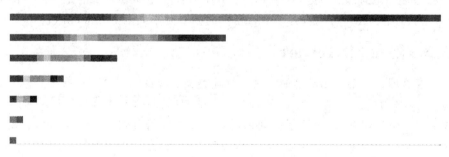

图 6-1 由 7 个 level 的 Image 组成的一维纹理

图 6-1 所示的一维纹理由 7 个 level 的 Image 共同组成,从上到下的 level 分别为 0~6。注意,为了便于观察,各个 level 的 Image 都被放大显示了。例如,最下面(level=6)的 Image 表示的是仅含有 1 个像素的 Image,这里为了便于观察,将其显示为一个小正方形,实际上其像素高度应为 1。如果需要将此纹理附着到 Framebuffer,需要指明要附着的是哪一个 Image。可以将同一个 Texture 对象的不同 Image 同时附着到不同的 Framebuffer 上。

6.3.2 附着 1D Array Texture

将多个一维纹理的内容排列成一个数组,视为一个整体,就是一维数组纹理(1D Array Texture),如图 6-2 所示。

图 6-2 由 3 层一维纹理的内容组成的一维数组纹理

1 个一维数组纹理对象仍旧只是 1 个 Texture 对象,而不是多个。在 Shader 中使用时,除了提供一维纹理所需的坐标,还需提供一个索引值,用于指定要用的纹理层。图 6-2 描述的是由 3 层一维纹理的内容组成的一维数组纹理,它共有 21 个 Image。如果需要将此纹理附着到 Framebuffer,需要指明要附着的是哪一层的哪一个 level 上的 Image。可以将各个 Image 同时附着到不同的 Framebuffer。

6.3.3 附着 2D Texture

二维纹理(2D Texture)是最常见的纹理类型。一个二维纹理对象可能包含多个 Image。这些 Image 按从大到小、边长除以 2 的规则存在,例如由 6 个 level 的 Image 组成的 Texture 如图 2-4 所示。如果需要将此纹理附着到 Framebuffer,需要指明要附着的是哪一个 level 上的 Image。可以将各个 Image 同时附着到不同的 Framebuffer。

6.3.4 附着 2D Array Texture

二维数组纹理(2D Array Texture)类型与一维数组纹理类似,它是将多个二维纹理的内容排列成一个数组,将其视为一个整体,如图 6-3 所示。

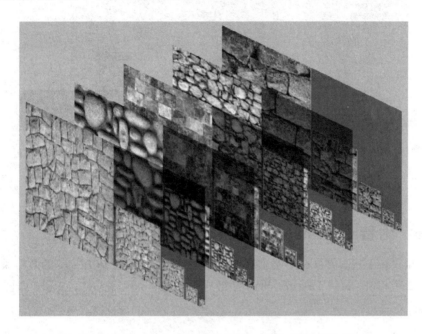

图 6-3　由 5 层二维纹理的内容组成的二维数组纹理

1 个二维数组纹理对象仍旧只是 1 个 Texture 对象,而不是多个。在 Shader 中使用时,除了提供二维纹理所需的坐标,还需提供一个索引值,用于指定要用的纹理层。图 6-3 描述的是由 5 层二维纹理的内容组成的二维数组纹理,它共有 25 个 Image。如果需要将此纹理附着到 Framebuffer,需要指明要附着的是哪一层的哪一个 level 上的 Image。可以将各个 Image 同时附着到不同的 Framebuffer。

6.3.5 附着 Cubemap Texture

Cubemap Texture 是由 6 个二维纹理的内容拼接成的立方体,如图 2-9 和图 6-4 所示。

注意,Cubemap Texture 也支持多个 level,每个 level 上都有 6 个 Image,而图 2-9 和图 6-4 中都仅画出了一个 level 上的 6 个 Image。

Cubemap Textures 里每个 level 都包含 6 个 Image。因此,1 个 Image 可以被 target 和 level 标识。然而在一些 OpenGL 函数里,1 个 level 里的各个面是用 layer 索引标识的。

图 6-4 Cubemap Texture 的某个 level 的 6 个 Image

6.3.6　附着 Cubemap Array Texture

　　Cubemap Array Texture 类似于 2D Array Texture，只是 Image 的数量乘以 6。一个由 2 个 Cubemap Texture 的内容组成的 Cubemap Array Texture 如图 6-5 所示。

图 6-5　(左)单独的 Cubemap　(右)由 2 个 Cubemap 成的 Cubemap Array Texture

　　因此一个 Image 由 layer(具体的应该是 layer-face)和 level 标识。

6.3.7 附着 3D Texture

3D Texture 的 1 个 mipmap level 被视作 2D Image 的集合,此集合的元素数量即为此 mipmap level 的 Z 坐标。Z 坐标的每个整数值都是一个单独的 2D 层(layer),因此 3D Texture 里的一个 Image 由 layer 和 mipmap level 共同标识。注意,3D Texture 的不同 level 的 Z 坐标数量是不同的。

6.3.8 附着 Rectangle Texture

Rectangle Texture 是一种特殊的二维纹理,其 UV 坐标范围不是标准化的[0,1],而是整数的[0,n],n 为其宽度或高度。

Rectangle Texture 只有 1 个 Image,因此直接用 level = 0 标识。

6.4 附着 Renderbuffer

创建一个 Texture 对象,然后将它的某个 Image 附着到 Framebuffer 上,这是一个繁冗的过程。如果只想使用自定义 Framebuffer 里的 Depth Buffer,那么就没必要为它的 Color Buffer 创建一个 Texture 对象。此时,只需简单地创建一个 Renderbuffer 对象,并将其附着到 Color Buffer 的位置即可。也就是说,在不需要读取 Color Buffer 位置上的内容时,可以让 Renderbuffer 起到一个占位符的作用,它不能像一般的 Texture 里的 Image 一样被 Shader 读取。

6.5 示例:Render to Texture

本例将绘制一个 Teapot 模型到 Framebuffer 的 Color Buffer 中,然后将此 Color Buffer 中的内容作为纹理贴到一个四边形上,效果如图 6-6 所示。

6.5.1 场 景

本例中,场景树如图 6-7 所示。

渲染本场景时,按照先序遍历的顺序,从上到下、从左到右地渲染整个场景。在执行到 Framebuffer Node 时,执行 Framebuffer 对象的初始化和绑定操作,然后以通常的方式渲染 Teapot Node,这样 Teapot 的渲染结果就保存到了 Framebuffer 里。最后,Rect Node 将 Framebuffer 对象保存的 Image 并渲染到 Rect 模型上。

注意,当 Teapot Node 渲染结束时,程序栈会回到 Framebuffer Node,这一刻,Framebuffer Node 会调用 RenderAfterChildren(..) 方法,来解绑 Framebuffer 对象。图 6-7 中的虚线描述了程序栈的执行路线。

OpenGL 简约笔记：用 C♯ 学面向对象的 OpenGL

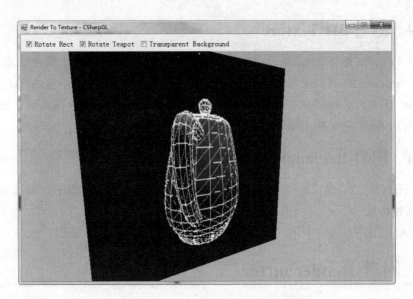

图 6-6　Render to Texture 效果图

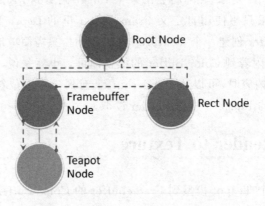

图 6-7　Render to Texture 的场景树

6.5.2　创建 Framebuffer 对象

本例中创建 Framebuffer 对象的代码如下：

```
// Create Framebuffer object with specified width and height.
private Framebuffer CreateFramebuffer(int width, int height)
{
    var framebuffer = new Framebuffer(width, height);
    framebuffer.Bind(); // start working on this fbo.
    {    // attach texture's first image to Color Buffer.
```

```
            var storage = new TexImageBitmap(width, height);// glTexImage(..., IntPtr.Zero);
            var texture = new Texture(storage, ...);
            texture.Initialize();
            uint location = 0; // first location of Color Buffer attachment points.
            int level = 0; // first level of specified texture.
            framebuffer.Attach(texture, location, level);
        }
        {   // attach renderbuffer to Depth Buffer.
            uint internalFormat = GL_DEPTH_COMPONENT24;
            var renderbuffer = new Renderbuffer(width, height, internalFormat);
            framebuffer.Attach(renderbuffer, AttachmentLocation.Depth);
        }
        framebuffer.CheckCompleteness(); // If something's wrong, we can't use this fbo.
        framebuffer.Unbind(); // stop working on this fbo.

        return framebuffer;
    }
```

Framebuffer 对象包含多个 Color Buffer 的附着点，因此可以同时附着多个 Color Buffer。但只有 1 个 Depth Buffer 和 1 个 Stencil Buffer 附着点，因此只能同时附着 1 个 Depth Buffer 和 1 个 Stencil Buffer。附着 Color Buffer 时，需要指明附着到第几个 Color Buffer 的附着点上。上述代码中的 location 就指明了要附着点的索引值，而 level 指明了要将 texture 中的哪个 Image 附着到 Framebuffer。

在使用此 Framebuffer 时，OpenGL 会把渲染结果填充到 texture 的指定 Image 里。由于本例不关心 Depth Buffer 中的内容，因此直接把一个 Renderbuffer 对象附着到 Depth Buffer 即可。

为了说明 Framebuffer 对象的多个 Color Buffer 附着点的用处，这里多为 Framebuffer 对象附着 3 个 Color Buffer，并在 Shader 中分别填入 Fragment 的 R、G 和 B 值，即让额外的 3 个 Texture 上的 Image 分别记录 Fragment 的 R、G 和 B 通道上的数据。附着额外 3 个 Color Buffer 的代码如下：

```
// Create Framebuffer object with specified width and height.
private Framebuffer CreateFramebuffer(int width, int height)
{
    var framebuffer = new Framebuffer(width, height);
    framebuffer.Bind(); // start working on this fbo.
    for (uint location = 0; location < 4; location++)
    {// attach texture's first image to Color Buffer.
        var storage = new TexImageBitmap(width, height);// glTexImage(..., IntPtr.Zero);
```

```
            var texture = new Texture(storage, ...);
            texture.Initialize();
            int level = 0; // first level of specified texture.
            framebuffer.Attach(texture, location, level);
    }
    {// attach renderbuffer to Depth Buffer...
    }
    framebuffer.SetDrawBuffers(GL_COLOR_ATTACHMENT0 + 0,
        GL_COLOR_ATTACHMENT0 + 1,
        GL_COLOR_ATTACHMENT0 + 2,
        GL_COLOR_ATTACHMENT0 + 3);
    framebuffer.CheckCompleteness(); // If something's wrong, we can't use this fbo.
    framebuffer.Unbind(); // stop working on this fbo.

    return framebuffer;
}
```

在一般的 Fragment Shader 中只有一个 out vec4 类型的变量用于记录 Fragment 的颜色值。这一 out 变量的值实际上就写入了 Framebuffer 的第一个 Color Buffer 上。为了同时也向额外的 3 个 Color Buffer 写入某些颜色值,需要在 Fragment Shader 中多定义 3 个 out 类型的变量,且声明其 location,代码如下:

```
// fragment shader.
#version 330 core

in vec3 passColor;

layout(location = 0) out vec4 outColor; // write to texture image at location 0.
layout(location = 1) out vec4 outRed;   // write to texture image at location 1.
layout(location = 2) out vec4 outGreen; // write to texture image at location 2.
layout(location = 3) out vec4 outBlue;  // write to texture image at location 3.

void main() {
    outColor = vec4(passColor, 1.0);
    outRed   = vec4(passColor.x, 0.0, 0.0, 1.0);
    outGreen = vec4(0.0, passColor.y, 0.0, 1.0);
    outBlue  = vec4(0.0, 0.0, passColor.z, 1.0);
}
```

注意,Fragment Shader 中的 location 与附着 Color Buffer 时的 location 变量是对应的,即在客户端代码 framebuffer.Attach(texture, location, level); 中的 location 为 t ($t = 0$、1、2 或 3)时,对应的是 Fragment Shader 中的 layout(location

= t）out vec4 ...这一变量。

如果将本例的 Fragment Shader 中的各项输出数据都导出为图片，则如图 6-8 所示。

图 6-8　（左上）location＝0　（右上）location＝1　（左下）location＝2　（右下）location＝3

另外，能够附着到 Color Buffer 上的 Texture 里的 Image 未必是 RGBA 这样的格式；与之对应的，Fragment Shader 中的 out 变量也未必是 vec4 类型。只要两者类型匹配且 location 相同，就能够将 Fragment Shader 的输出值写入 Color Buffer 中。这一灵活的配置能力使得 Fragment Shader 能够实现非常多的效果。

本例中其他要点在本书之前的章节已有所介绍，不再重复。读者可在本书网络资料上的 Demos\RenderToTexture 项目中查看完整源代码。

6.6　总　结

本章介绍了 Framebuffer 的结构和用法。
● Framebuffer 对象至少包含什么？
一个 Color Buffer 和一个 Depth Buffer。
● 将 Texture 附着到 Framebuffer 的本质？
将 Texture 里的某个 Image 附着到 Framebuffer。
● 如何使用 Framebuffer？
创建 Framebuffer 对象→附着 Image 或 Renderbuffer→检查完整性→绑定→渲染（附着的 Image 收获数据）→解绑→使用附着的 Image。

6.7 问题

带着问题实践是学习 OpenGL 的最快方式。这里给读者提出几个问题，作为抛砖引玉之用。

1. 请尝试使用各种类型的 Texture，将其包含的 Image 附着到自定义的 Framebuffer 对象。

2. 请尝试将本章示例中的矩形 Node 替换为波浪形状的 Node，观察类似飘动的旗子的效果。

第 7 章

阴 影

学完本章之后,读者将能:
- 使用 Shadow Mapping 技术为场景中的物体添加阴影(Shadow);
- 使用 Shadow Volume 技术为场景中的物体添加阴影。

7.1 Shadow Mapping

在 Blinn-Phong 光照模型的基础上,如果能够为模型添加阴影效果,将更加提高场景的真实感,如图 7-1 所示。

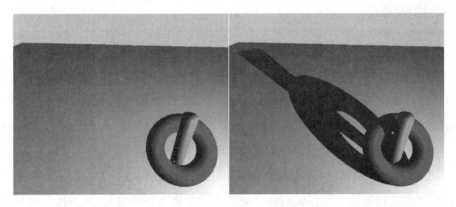

图 7-1 (左)没有阴影的模型 (右)用 Shadow Mapping 方法为模型创建阴影

图 7-1 展示了用 Shadow Mapping 方法为一个圆柱模型和一个圆环模型创建阴影后的效果。这 2 个模型的影子不仅投射到地面上,而且圆柱上的部分影子也投射到了圆环上。可以看到,渲染了阴影的场景生动地展示了模型的位置,更具有空间立体感。读者可以在本书网络资料上的 c07d00_ShadowMapping 项目中找到此示例的完整代码。

为了实现 Shadow Mapping,需要使用 OpenGL Pipeline 中尚未介绍的功能——深度测试。因此,这里先介绍相关的基础知识点。

7.1.1 逐 Fragment 的测试和操作

在 5.5.1 小节中,介绍在同时使用多个光源时提到,在 Pipeline 的 Fragment Shader 阶段之后,还要对每个 Fragment 做多项测试和操作。只有成功完成全部测试和操作的片段才能显示到画布上,成为像素。这些测试和操作的先后顺序如图 7-2 所示。

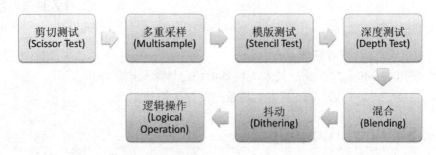

图 7-2 在 Fragment Shader 之后依次进行的测试和操作

本节逐一简要介绍这些测试或操作。

1. 剪切测试

剪切测试可以将所有的渲染操作都局限在画布上的一个矩形区域内,矩形区域外的内容不会改变。这就像一个挡板,只有在矩形范围内的操作才会继续,矩形范围外的操作都被这一步挡住了,不能再继续,从而取消了。显然这一操作是在 Window Space 内发生的。使用如下 OpenGL 函数可以设定此矩形的位置和范围:

```
// restrict renderingto specified area.
void glScissor(int x, int y, int width, int height);
```

需要启用剪切测试时,只需调用 glEnable(GL_SCISSOR_TEST);。剪切测试开启后,只有在设定的矩形范围内的渲染操作才能影响最终结果,范围外的内容不会发生改变。窗口的清除操作 glClear(..) 也会受到此限制。

2. 多重采样

多重采样消耗额外的资源,能够使渲染结果更加平滑,由于暂时不使用此功能,在此不做介绍。

3. 模版测试

颜色缓存(Color Buffer)、深度缓存(Depth Buller)和模板缓存(Stencil Buffer)都是 Framebuffer 对象中的缓存对象。类似深度缓存记录了深度测试的结果,模板缓存记录了模板测试的结果。

模版测试最有用的一个功能,也许就是通过掩码在窗口上构成一个不规则的区域,然后阻止在此区域内进行任何渲染操作。在本书网络资料上的 Demos\StencilTest 项目,展示了如何利用模版测试给指定的 Cube 模型添加外包围框的功能,如图 7-3 所示。

图 7-3　利用模版测试添加包围框

实现此功能的大体思路:首先启用模版测试,然后按照常规方式渲染 Cube 模型,此时,Cube 模型影响到的 Fragment 所在的位置都会被记录到模板缓存中;然后,稍微放大 Cube,并且使用包围框需要的颜色,再次渲染此 Cube,与此同时,启用模版测试,并确保上一次渲染占用的区域内禁止修改颜色缓存。这样,包围框只会渲染到 Cube 放大的部分,而正常渲染的 Cube 也能够保持下来。

整体来看,使用模版测试一般需要 2 个步骤:第一步,执行渲染,同时让模版缓存记录需要禁止修改的区域范围(Area);第二步,执行渲染,同时避免修改 Area 内的内容。第一步的目的是收获模板缓存的状态,不是将图形渲染出来。第二步的目的才是根据模板缓存的状态,实现一些特殊的渲染效果。

后续章节将详细介绍模版测试的用法。

4. 深度测试

每个 Fragment 都是由模型的顶点通过插值形成的。每个 Fragment 到 Camera 都有一定的距离,即深度值。当前绑定的 Framebuffer 对象的深度缓存记录了每个像素的深度值。如果新 Fragment 的深度值小于 Framebuffer 中记录的当前深度值(即新 Fragment 更靠近 Camera),那么新 Fragment 的颜色就可以覆盖 Framebuffer 在这一像素位置的颜色。这样就实现了近处物体遮挡远处物体的效果,这是深度测试的典型应用。

当然,用户可以通过 glDepthFunc(..) 来指定在何种情形下用新 Fragment 代替原有的像素颜色。

5. 混 合

如果一个 Fragment 通过了前面的测试和操作,就可以与当前绑定的 Framebuffer 对象里包含的颜色缓存中的颜色进行合并了。默认的合并方式,就是新颜色覆盖已有颜色。如果启用了混合功能,就可以将新颜色与已有颜色按照某种方式叠加,从而实现各种绚丽或者诡异的效果。大多情况下,混合时需要用到描述颜色值的第四个分量 alpha,即 vec4(r, g, b, a) 中的 a。即使用户没有显式地设置 alpha,OpenGL 中的颜色也会带有 alpha 值。alpha 不像 rgb 那样能够被直观地看到,它只能用于描述透明度,使用 alpha 可以实现各种半透明物体的效果,例如有色玻璃。

注意,按照从远到近的顺序渲染,并且使用 glBendFunc(GL_SRC_ALPHA, GL_ONE_MINUS_DST_ALPHA);这样的设置,才能观察到正确的半透明渲染效果。否则,应当使用一些高级的技巧(例如与顺序无关的透明算法 Order-Independent-Transparency)。后续章节将介绍几种实现的方法。

6. 抖 动

对于颜色位数较小的系统,可以通过对图像中的颜色进行抖动来提升颜色的分辨率。抖动操作是与硬件相关的。OpenGL 中只允许开启或关闭此操作。如果机器上的分辨率已经很高,那么开启抖动可能不会产生任何效果。

7. 逻辑操作

逻辑操作类似混合功能,它也作用于颜色缓存中当前的颜色和新 Fragment 中的颜色,通过对两者进行某种位操作,将得到的结果作为新颜色赋值给颜色缓存,这个过程的实现代价对于硬件来说是十分低廉的。

借助逻辑操作,可以实现将指定的模型反色显示。这一功能可以用作高亮或突出显示模型的某部分。请读者在本书网络资料中打开 Demos\LogicOperation 项目,运行此项目,可以看到使用逻辑操作后使得模型显示出反色的效果,如图 7-4 所示。

图 7-4 使用逻辑操作,使得最近的 Cube 以反色显示

使用 OpenGL 函数 glLogicOp(uint opcode); 即可设置需要的逻辑操作,使用 glEnable(GL_COLOR_LOGIC_OP); 或 glDisable(GL_COLOR_LOGIC_OP); 即可启用或关闭逻辑操作。

至此,OpenGL Pipeline 的概念如图 7-5 所示。

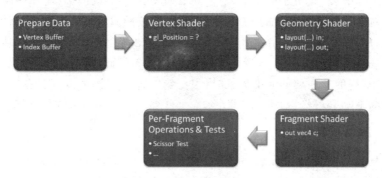

图 7-5　进一步细化的 OpenGL Pipeline

7.1.2　Shadow Mapping 核心思想

Shadow Mapping 的主要思想:给定光源 L,在它的一条光线穿过的所有 Fragment 中,距离 L 最近的(第一个被穿过的)那个 Fragment 能够被光源 L 照射到,其他 Fragment 处于被遮挡的状态,即位于阴影中,如图 7-6 所示。

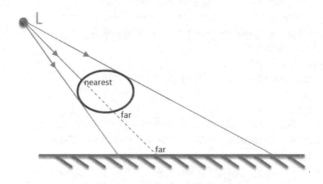

图 7-6　Shadow Mapping 原理

如图 7-6 所示,光源 L 照射到椭圆形物体和平整的地面上。对于 L 的一条光线,其与椭圆形物体的接触点有 2 个,与地面的接触点有 1 个。L 最先照射到的接触点为最近点(nearest),此处应按常规的光照模型渲染,而且另外 2 个接触点都被视为远点(far),都应用 discard 或其他方式来取消光源 L 对这些点的光照计算。那么,如何得到各个像素上的 Fragment 到 L 的距离呢?

注意,在 Camera 的位置观察场景时,每个 Fragment 的深度值都在深度测试时出现过,它们参与了深度测试过程,距离 Camera 最近的那个 Fragment 被保留了下来。这意味着,如果从光源 L 的位置观察场景,那么距离 L 最近的 Fragment 的深度值就会记录到深度缓存中。借助自定义的 Framebuffer 对象,可以将此深度缓存保留下来。如果在使用 Blinn-Phong 渲染场景时,将距离大于深度缓存中对应像素的深度值的 Fragment 的光照计算忽略掉,那么此处就会形成阴影;而那些距离等于深度缓存中对应像素的深度值的 Fragment,则仍旧按照原有方式参与光照计算。

因此,用 Shadow Mapping 渲染阴影需要调用 2 遍 Pipeline:第一遍,从光源 L 的位置观察场景(即将 Camera 放置于光源 L 的位置),将此时的深度缓存保存起来;第二遍,在按常规的光照模型渲染场景时,根据第一遍中得到的深度缓存信息,忽略处于阴影中的 Fragment。

7.1.3　获取深度图

虽然从程序实现的角度看,能够直接用默认 Framebuffer 对象承接第一遍的 Pipeline 的输出结果,并用于第二遍的 Pipeline,但是程序比较复杂,且效率较低。因此,这里有必要创建一个自定义的 Framebuffer 对象,并且用一个二维纹理对象的 Image 充当深度缓存。创建这个 Framebuffer 对象的代码如下:

```csharp
// Creates framebuffer for shadow mapping.
Framebuffer CreateFramebuffer(int width, int height)
{
    var framebuffer = new Framebuffer(width, height); // framebuffer object.
    framebuffer.Bind(); // all operations will happen to this framebuffer object.
    // Creates texture object that will keep the depth information from light's view.
    var texture = new Texture(new TexImage2D(GL_DEPTH_COMPONENT32, width, height, GL_DEPTH_COMPONENT, GL_FLOAT),
        // default filter mode.
        new TexParameteri(GL_TEXTURE_MIN_FILTER, GL_LINEAR),
        new TexParameteri(GL_TEXTURE_MAG_FILTER, GL_LINEAR),
        // depth comparison mode.
        new TexParameteri(GL_TEXTURE_COMPARE_MODE, GL_COMPARE_REF_TO_TEXTURE),
        new TexParameteri(GL_TEXTURE_COMPARE_FUNC, GL_LEQUAL),
        // clamp mode.
        new TexParameteri(GL_TEXTURE_WRAP_S, GL_CLAMP_TO_EDGE),
        new TexParameteri(GL_TEXTURE_WRAP_T, GL_CLAMP_TO_EDGE));
    texture.Initialize();
    // Texture's image will work as this framebuffer object's Depth Buffer.
    framebuffer.Attach(texture, AttachmentLocation.Depth);

    framebuffer.CheckCompleteness(); // check if this framebuffer is ready to work.
    framebuffer.Unbind();

    return framebuffer;
}
```

注意，这个用于记录深度缓存的 Framebuffer 对象，只有一个深度缓存，没有颜色缓存。也就是说，这个 Framebuffer 对象不能记录第一遍渲染产生的图像，这种情形的 Framebuffer 是可以合法存在的。由于第一遍渲染的目的仅仅是获取深度缓存中的数据，所以颜色缓存实际上是不需要的。这里不添加颜色缓存，避免了浪费。

另外，由于 Fragment Shader 的处理结果会写入颜色缓存，而此时并不存在颜色缓存，因此第一遍渲染时也不需要为 Shader Program 指定 Fragment Shader。也就是说，只有 Vertex Shader 而没有 Fragment Shader 的情形也是可以合法存在的。此时，第一遍渲染使用的 Pipeline 中被激活的部分如图 7-7 所示。

图 7-7　Shadow Mapping 第一遍渲染激活的 Pipeline

图 7-8 展示了在光源处于不同位置时的深度缓存的内容。

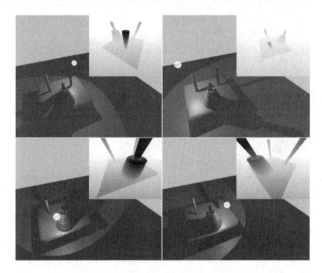

图 7-8　光源处于不同位置时的深度缓存

图中的光源使用了聚光灯，小球代表光源的位置。此图分为 4 个部分，分别展示了光源位于不同的位置时，深度缓存的内容和最终的渲染效果。深度缓存中的白色部分表示远处，越接近黑色说明越靠近光源。例如在左下方的部分，光源靠近圆盘模型，因此在深度缓存中可以明显看到圆盘的颜色深，而远处柱子的颜色浅。

7.1.4　渲染有阴影的场景

依据从第一遍的渲染过程中得到的深度信息，忽略阴影部分的光照计算，就实现了第二遍渲染过程。注意，阴影部分不参与 Blinn-Phong 的漫反射光和镜面反射光的计算，但是仍然要参与环境光的计算。如果场景中有多个光源，阴影部分也可能会参与其他光源的计算。

1. Vertex Shader

从光源位置观察时，顶点位置最终投射到深度缓存中。深度缓存中保存的位置，是位于 Clip Space 中的位置。在第二遍渲染时，必须再次计算此位置，这样才能向保存深度信息的 Texture 对象提供此顶点的 UV 坐标。第二遍的 Vertex Shader 代码如下：

```
// vertex shader.
#version 150

in vec3 inPosition;
in vec3 inNormal;

// Declare an interface block.
outVertex {
    vec3 position;
    vec3 normal;
    vec4 shadowCoord;
} v;

uniform mat4 mvpMat;
uniform mat4 modelMat;
uniform mat4 normalMat;// transpose(inverse(modelMat));
uniform mat4 shadowMatrix;

void main() {
    gl_Position = mvpMat * vec4(inPosition, 1.0);
    vec4 worldPos = modelMat * vec4(inPosition, 1.0);
    v.position = worldPos.xyz;
    v.normal = (normalMat * vec4(inNormal, 0)).xyz;
    v.shadowCoord = shadowMatrix * worldPos; // position in clip space + vec3(0.5)
}
```

由于 Shader 中对每个顶点都计算一次 transpose(inverse(modelMat)); 是比较费时的，所以这里直接在应用程序代码中一次性计算出结果并传递给 Shader。

shadowMatrix 的作用是把顶点变换到 Clip Space 后再变换到描述纹理坐标的空间。这样得到的就是顶点在深度缓存 Texture 里的 UV 坐标，可以直接用于检索 Texture 中的数据。这一矩阵可以用下述方式直接获得：

```
mat4 GetShadowMatrix(mat4 lightProjection, mat4 lightView)
{
    mat4 bias = mat4.identity();
    bias = glm.translate(bias, new vec3(0.5f, 0.5f, 0.5f));
    bias = glm.scale(bias, new vec3(0.5f, 0.5f, 0.5f));
    mat4 shadowMatrix = bias * lightProjection * lightView;

    return shadowMatrix;
}
```

其中 lightProjection 和 lightView 是将 Camera 放到光源的位置上时的 Projection 矩阵和 View 矩阵。其他部分都属于 Blinn-Phong 模型的代码,不再赘述。

2. Fragment Shader

在 Fragment Shader 中,除了要判断此 Fragment 是否处于阴影中,其他均与常规的 Blinn-Phong 代码相同。Fragment Shader 代码如下:

```
//fragment shader.
#version 150

uniform sampler2DShadow texShadow;

inVertex {
    vec3 position;
    vec3 normal;
    vec4 shadowCoord;
}v;

out vec4 outColor;

void main() {
    float diffuse = ...
    float specular = ...
    float f = textureProj(texShadow, v.shadowCoord);

    outColor = vec4(f * diffuse * light.diffuse * material.diffuse + f * specular
              * light.specular * material.specular, 1.0);
}
```

Vertex 承接了来自 Vertex Shader 的顶点信息。纹理的采样器类型不是常规的 sampler2D,而是 sampler2DShadow ,这一类型配合 textureProj(..) 函数能够在阴影边缘处给出平滑变动的返回值,返回值范围为[0, 1]。当 Fragment 完全位于阴影中，f 的值为 0；当 Fragment 完全位于光照中，f 的值为 1。因此，只需用返回值 f 与光照模型 Blinn-Phong 的常规计算结果相乘，即可得到带有阴影效果的光照结果。而且还能在光与影的交界处呈现比较平滑的过渡效果。

漫反射强度 diffuse 和镜面反射强度 specular 的计算方式与常规的光照模型相同。

7.2　Shadow Volume

上一节提到在 Per-Fragment Operations & Tests 阶段，有一个步骤是模版测试。依靠这一步骤，不仅可以实现渲染模型的包围框这样的实用功能，还能创造出一种渲染阴影的算法，即 Shadow Volume 算法。

用 Shadow Mapping 方法得到的阴影，在贴近观察时，会看到细微的锯齿。这是因为深度缓存受阴影贴图分辨率的限制，不可能完全精确地描述贴近观察时的各个 Fragment，但 Shadow Volume 方法得到的阴影是没有这样的锯齿问题的，如图 7-9 所示。

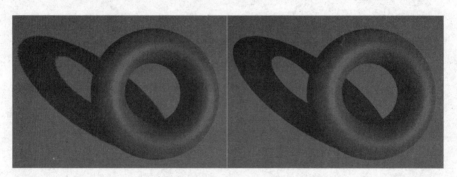

图 7-9　（左）Shadow Mapping 的阴影有锯齿　（右）Shadow Volume 的阴影更平滑

如图 7-9 所示，左侧的 Shadow Mapping 方法得到的阴影呈现出比较明显的锯齿，而右侧的 Shadow Volume 方法得到的阴影则十分平滑，这是由 Shadow Volume 的实现机制决定的。其机制概括起来就是，根据光源位置（或方向）和模型位置，实时动态地生成一个不规则的包围盒模型，使之恰好将阴影部分包裹起来（遮挡住），进而保证包围盒内部的 Fragment 不参与光照计算。图 7-10 展示了图 7-9（右图）中未被画出的包围盒。

注意，这个包围盒是实时动态生成的，它会随着光源的位置（或方向）或者模型自身的变化而变化，这个包围盒是可以延伸到无限远的，这样才能正确地渲染出阴影。

那么有两个主要问题：首先，如何动态地生成这个包围盒，而且能够覆盖无限远的范围；其次，如何根据包围盒判断 Fragment 是否参与光照计算。

为了便于理解问题，这里假设探讨的模型都是由三角形网格拼接组成的。例如

阴 影 7

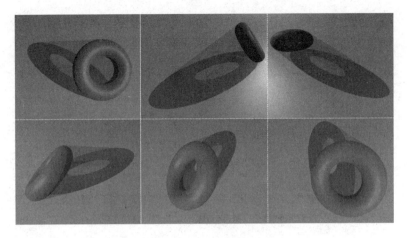

图 7-10 从 6 个视角观察包围盒

对于圆环模型和 Cube 模型，其三角形网格结构如图 7-11 所示。

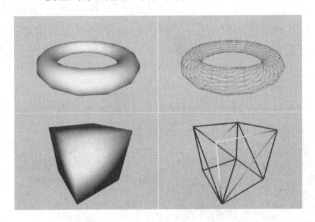

图 7-11 三角形网格组成的模型

可以看到，圆环模型是由非常多的三角形网格拼接而成的，这利于观察光照效果的真实感。Cube 模型则仅仅由 12 个三角形拼接而成，这利于检测程序的正确性。请读者在本书网络资料中找到任意一个可以渲染圆环模型的项目，为其添加 PolygonModeSwitch 开关，近距离观察模型的三角形网格。

7.2.1 包围盒

要找到一个模型在光源照射下的包围盒，首先要找到在光源照射下的外围轮廓（Outline）。轮廓的一侧都是能被光源照射到的三角形，另一侧都是不能被光源照射到的三角形（即处于阴影中），如图 7-12 所示。

在图 7-12 中，左上方有一个聚光灯光源。左图的白线勾勒出如图所示的轮廓线，中间的图为模型本身，右图为隐藏了模型的轮廓线全貌。

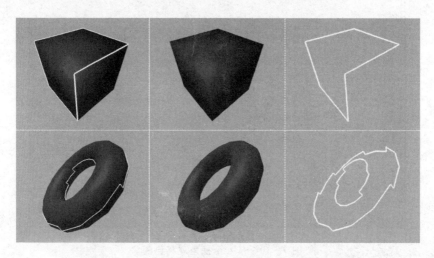

图 7-12 （左）模型＋轮廓线 （中）模型 （右）轮廓线

轮廓是由一条条线段组成的。对于组成线段的每个顶点,沿着从光源位置到顶点位置的方向,无限地延伸出去,就是要找的包围盒的侧面。接着加上轮廓中朝向光源的一侧,最后把无限远处封口,就得到了一个完整的包围盒。完整的包围盒如图 7-13 所示。

图 7-13 观察 Cube 模型的完整的包围盒

完整的包围盒从模型的位置无限地延伸到了场景的边缘。图 7-13 左边为地面遮挡了一部分包围盒的情形,可见包围盒与地面的交界处就是阴影所在,图右边为隐藏了地面后显示出的更完整的包围盒。可以看到包围盒在最远处仍然呈现出模型轮廓的形状,且包围盒的近端和远端保持着对应关系。

注意,包围盒是一个完全封闭的盒子,其法线全部指向外侧。这是在构造包围盒时特意设计的,因为这样才能在后续的判断过程中找到阴影。

7.2.2 动态生成包围盒

如何判断哪个顶点是在轮廓线上的呢? 观察图 7-14。

如图 7-14 所示,两个三角形的交界处,是一条共享的线段 AB。三角形 ABC 面

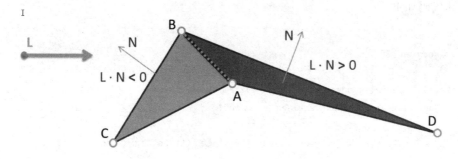

图 7-14　光线 L 照射到 2 个三角形上

朝光源 L，能够被照到，另一个三角形 ABD 则背朝光源 L，即处于阴影中。那么这条共享的线段 AB 就应当成为轮廓线的一部分。如果两个三角形同时面朝光源或同时背朝光源，那么它们之间的共享线段就不需要算到轮廓线里。要判断一个三角形是否正面朝光源，只需将光源的方向向量 L 与三角形面的法线向量 N 做 dot 乘法，如果结果为负数，说明正面朝光源；如果结果为正数，说明背朝光源。

注意，这里使用的是三角形面的法线向量。Vertex Shader 只能处理单个的顶点。要处理三角形面这样的对象，就要使用 Geometry Shader。为了得知一个三角形与周围哪些三角形有共享边，需要使用 GL_TRIANGLES_ADJACENCY 模式来渲染模型。这样的模型有一个特点，即每个三角形都包含了其周边三角形的信息，如图 7-15 所示。

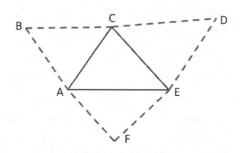

图 7-15　GL_TRIANGLES_ADJACENCY 模式的图元

图 7-15 展示了一个三角形网格模型的一部分（由 6 个顶点组成的 4 个三角形）。其中三角形 ACE 是 Geometry Shader 要处理的一个三角形，而三角形 ABC、CDE 和 EFA 是与三角形 ACE 有共享边的三个三角形。这三个三角形被称为三角形 ACE 的邻接三角形。按照 ABCDEF 的顺序向 Geometry Shader 依次传入这 6 个顶点，即可找到哪些边构成了模型的轮廓线。一般的，模型数据中是不包含邻接三角形信息的，这需要在加载模型后额外计算。

为实现 Shadow Volume 算法，找轮廓线的任务是在 Geometry Shader 中完成的，其代码如下：

```glsl
// geometry shader.
#version 330

layout (triangles_adjacency) in; // six vertices in
layout (line_strip, max_vertices = 6) out;

in vec3 inPositions[]; // an array of 6 vertices (triangle with adjacency)

uniform vec3 lightPos; // point light's position.
uniform mat4 vpMat; // projection * view
uniform mat4 modelMat;

// Emit a line using aline strip
void EmitOutline(vec3 startVertex, vec3 endVertex)
{
    gl_Position = vpMat * vec4(startVertex, 1.0);
    EmitVertex(); // first vertex of the line.

    gl_Position = vpMat * vec4(endVertex, 1.0);
    EmitVertex(); // second vertex of the line.

    EndPrimitive(); // the line is complete.
}

void main()
{
    vec3 worldPos[6]; // vertexes' positions in world space.
    worldPos[0] = vec3(modelMat * vec4(inPositions[0], 1.0));
    worldPos[1] = vec3(modelMat * vec4(inPositions[1], 1.0));
    worldPos[2] = vec3(modelMat * vec4(inPositions[2], 1.0));
    worldPos[3] = vec3(modelMat * vec4(inPositions[3], 1.0));
    worldPos[4] = vec3(modelMat * vec4(inPositions[4], 1.0));
    worldPos[5] = vec3(modelMat * vec4(inPositions[5], 1.0));
    // edges.
    vec3 e02 = worldPos[2] - worldPos[0];
    vec3 e04 = worldPos[4] - worldPos[0];
    vec3 e01 = worldPos[1] - worldPos[0];
    vec3 e23 = worldPos[3] - worldPos[2];
    vec3 e24 = worldPos[4] - worldPos[2];
    vec3 e05 = worldPos[5] - worldPos[0];
```

```
        vec3 normal = normalize(cross(e02, e04)); // normal of triangle face.
        vec3 lightDir = normalize(lightPos - worldPos[0]); // direction from vertex to
light.

        // Handle only light facing triangles
        if (dot(normal, lightDir) > 0) {
            normal = cross(e01, e02);
            if (dot(normal, lightDir) <= 0) {
                vec3 startVertex = worldPos[0];
                vec3 endVertex = worldPos[2];
                EmitOutline(startVertex, endVertex);
            }

            normal = cross(e23, e24);
            lightDir = normalize(lightPos - worldPos[2]);
            if (dot(normal, lightDir) <= 0) {
                vec3 startVertex = worldPos[2];
                vec3 endVertex = worldPos[4];
                EmitOutline(startVertex, endVertex);
            }

            normal = cross(e04, e05);
            lightDir = normalize(lightPos - worldPos[4]);
            if (dot(normal, lightDir) <= 0) {
                vec3 startVertex = worldPos[4];
                vec3 endVertex = worldPos[0];
                EmitOutline(startVertex, endVertex);
            }
        }
    }
```

注意，代码中的 lightDir 变量指的是从顶点到光源位置的向量，而图 7-14 中的光源方向向量 L 是从光源到顶点的向量，两者是相反的。因此在代码中面朝光源的三角形面的法向量 N 与 lightDir 的 dot 结果为正数。

有了轮廓线，将轮廓线的每一条线段都延伸出去，分别形成一个四边形，就构成了包围盒的侧面。所有面朝光源的三角形，就构成了包围盒的近顶。沿着包围盒侧面的方向，把各个近顶面分别推向无限远处，并且翻转朝向，就构成了包围盒的远底。

只需在生成轮廓线代码的基础上稍作修改，即可得到动态生成包围盒的 Geometry Shader，代码如下：

```glsl
// geometry shader.
#version 330

layout (triangles_adjacency) in; // six vertices in
layout (triangle_strip, max_vertices = 18) out;// 4 per quad * 3 triangle vertices + 6 for near/far caps

in vec3 inPositions[]; // an array of 6 vertices (triangle with adjacency)

uniform vec3 lightPos; // point light's position.
uniform mat4 vpMat;
uniform mat4 modelMat;

float EPSILON = 0.001;

// Emit a quad using a triangle strip
void EmitQuad(vec3 startVertex, vec3 endVertex)
{
    vec3 lightDir;
    lightDir = normalize(startVertex - lightPos);
    // Vertex #1: the starting vertex (just a tiny bit below the original edge)
    gl_Position = vpMat * vec4((startVertex + lightDir * EPSILON), 1.0);
    EmitVertex();
    // Vertex #2: the starting vertex projected to infinity
    gl_Position = vpMat * vec4(lightDir, 0.0);
    EmitVertex();

    lightDir = normalize(endVertex - lightPos);
    // Vertex #3: the ending vertex (just a tiny bit below the original edge)
    gl_Position = vpMat * vec4((endVertex + lightDir * EPSILON), 1.0);
    EmitVertex();
    // Vertex #4: the ending vertex projected to infinity
    gl_Position = vpMat * vec4(lightDir, 0.0);
    EmitVertex();

    EndPrimitive();
}

void main()
{
    vec3 worldPos[6]; // vertexes' position in world space.
    worldPos[0] = vec3(modelMat * vec4(inPositions[0], 1.0));
```

```glsl
worldPos[1] = vec3(modelMat * vec4(inPositions[1], 1.0));
worldPos[2] = vec3(modelMat * vec4(inPositions[2], 1.0));
worldPos[3] = vec3(modelMat * vec4(inPositions[3], 1.0));
worldPos[4] = vec3(modelMat * vec4(inPositions[4], 1.0));
worldPos[5] = vec3(modelMat * vec4(inPositions[5], 1.0));
vec3 e02 = worldPos[2] - worldPos[0];
vec3 e04 = worldPos[4] - worldPos[0];
vec3 e01 = worldPos[1] - worldPos[0];
vec3 e23 = worldPos[3] - worldPos[2];
vec3 e24 = worldPos[4] - worldPos[2];
vec3 e05 = worldPos[5] - worldPos[0];

vec3 normal = normalize(cross(e02,e04));
vec3 lightDir;
lightDir = normalize(lightPos - worldPos[0]);

// Handle only light facing triangles
if (dot(normal, lightDir) > 0) {
    normal = cross(e01, e02);
    if (dot(normal, lightDir) <= 0) {
        EmitQuad(worldPos[0], worldPos[2]);
    }

    normal = cross(e23, e24);
    lightDir = normalize(lightPos - worldPos[2]);
    if (dot(normal, lightDir) <= 0) {
        EmitQuad(worldPos[2], worldPos[4]);
    }

    normal = cross(e04, e05);
    lightDir = normalize(lightPos - worldPos[4]);
    if (dot(normal, lightDir) <= 0) {
        EmitQuad(worldPos[4], worldPos[0]);
    }

    // render the front(near) cap
    lightDir = (normalize(worldPos[0] - lightPos));
    gl_Position = vpMat * vec4((worldPos[0] + lightDir * EPSILON), 1.0);
    EmitVertex();
    lightDir = (normalize(worldPos[2] - lightPos));
    gl_Position = vpMat * vec4((worldPos[2] + lightDir * EPSILON), 1.0);
```

```
                EmitVertex();
                lightDir = (normalize(worldPos[4] - lightPos));
                gl_Position = vpMat * vec4((worldPos[4] + lightDir * EPSILON), 1.0);
                EmitVertex();
                EndPrimitive();

                // render the back(far) cap
                lightDir = worldPos[0] - lightPos;
                gl_Position = vpMat * vec4(lightDir, 0.0);
                EmitVertex();
                lightDir = worldPos[4] - lightPos;
                gl_Position = vpMat * vec4(lightDir, 0.0);
                EmitVertex();
                lightDir = worldPos[2] - lightPos;
                gl_Position = vpMat * vec4(lightDir, 0.0);
                EmitVertex();
                EndPrimitive();
            }
        }
```

上文提到,包围盒的远底面是位于无限远处的,这是数学意义上的描述。具体到 OpenGL,其实并不需要描述一个无限远的顶点,只需要找到此顶点投影到近裁剪面上的投影位置即可。简单来说,只需将从光源到轮廓线上的顶点的向量作为 xyz 坐标,将 0 作为 w 坐标,即可得到此投影位置。

从数学上理解此问题需要一些晦涩的推导过程,这里从 OpenGL Pipeline 的角度来理解即可。一般来说,OpenGL 描述顶点位置都是用 vec4(x, y, z, 1)。在 Pipeline 从 Clip Space 到 NDS Space 的变换过程中,会将所有顶点的 xyz 坐标都除以 w,所以 vec4(x, y, z, w)、vec4(x/w, y/w, z/w, 1) 和 vec4(nx, ny, nz, nw) 描述的都是同一个位置。试想,如果保持 xyz 的值不变,而不断地减小 w 的值,那么这个坐标描述的位置就会越来越远;当 w 不断地逼近到 0 时,这个坐标描述的就是一个无限远的位置。那么,沿着光源 L 到顶点的方向,走到无限远的那个位置,只能是 vec4(lightDir, 0)。

7.2.3 判　断

使用 Stencil Buffer 和 Depth Buffer 实现阴影过程的伪代码如下:

```
void ShadowVolume(Scene scene, ..)
{
    // Render depth info into Depth Buffer.
    RenderDepthInfo(scene, ..);

    glEnable(GL_STENCIL_TEST); // enable stencil test.
    glClear(GL_STENCIL_BUFFER_BIT); // Clear Stencil Buffer.
    // Extrude shadow volume and save shadow info into Stencil Buffer.
    {
        glDepthMask(false); // Disable writing to Depth Buffer.
        glColorMask(false, false, false, false); // Disable writing to Color Buffer.
        glDisable(GL_CULL_FACE); // Disable culling face.

        // Set up stencil function and operations.
        glStencilFunc(GL_ALWAYS , 0, 0xFF);
        glStencilOpSeparate(GL_BACK, GL_KEEP, GL_INCR_WRAP , GL_KEEP);
        glStencilOpSeparate(GL_FRONT, GL_KEEP, GL_DECR_WRAP , GL_KEEP);

        // Extrude shadow volume.
        // Shadow info will be saved into Stencil Buffer automatically
        // according to glStencilOp...
        Extrude(scene, ..);

        // Reset OpenGL switches.
        glEnable(GL_CULL_FACE);
        glColorMask(true, true, true, true);
        glDepthMask(true);
    }
    // Render the scene under the light with shadow.
    {
        // Set up stencil function and operations.
        glStencilFunc(GL_EQUAL , 0x0, 0xFF);
        glStencilOp(GL_KEEP, GL_KEEP, GL_KEEP);

        // light the scene up.
        RenderUnderLight(scene, ..);
    }
    glDisable (GL_STENCIL_TEST); // disable stencil test.
}
```

Shadow Volume 由 3 遍渲染完成。

第一遍渲染时,在不考虑阴影的前提下正常渲染场景。此时,Depth Buffer 填充了正常的深度信息。这一次渲染的目的是准备好这一深度缓存,渲染的颜色并不重要。因此可以用最简单的 Fragment Shader,甚至不使用 Fragment Shader。

第二遍渲染前,启用模板测试,并按如下方式设置模板测试的函数和操作:

```
//Always pass stencil test.
glStencilFunc(GL_ALWAYS, 0, 0xFF);
// If depth test fails for back face, increase value in Stencil Buffer.
glStencilOpSeparate(GL_BACK, GL_KEEP, GL_INCR_WRAP, GL_KEEP);
// If depth test fails for front face, decrease value in Stencil Buffer.
glStencilOpSeparate(GL_FRONT, GL_KEEP, GL_DECR_WRAP, GL_KEEP);
```

这里设置模版测试对于每个像素都是通过的,且通过后将对应像素位置的模板缓存值设置为 0。当模板测试完成后,对于包围盒背面的 Fragment,如果它的深度测试失败,那么模版缓存的值加 1;对于包围盒正面的 Fragment,如果它的深度测试失败,那么模板缓存的值减 1。

这样设置的结果是,对于被包围盒包裹起来的模型所在的像素位置,包围盒背面的深度测试会失败,所以此处的模板缓存值加 1;包围盒正面的深度测试会成功,所以对模板缓存无影响。对于比包围盒更靠近 Camera 的模型所在的像素位置,包围盒的正面和背面的深度测试都会失败,所以此处的模板缓存值加 1 又减 1,保持为 0。对于比包围盒更远离 Camera 的模型所在的像素位置,包围盒的正面和背面的深度测试都会成功,所以此处的模板缓存值保持不变,即为 0。而在包围盒涉及不到的像素位置,模板缓存也保持不变,即为 0。

也就是说,只有被包围盒包裹起来的模型所在的像素位置,其对应的模板缓存值是大于 0 的,其他位置的模板缓存值都保持为 0。而被包围盒包裹起来的模型,恰恰是位于阴影中的。所以第二遍渲染时,只渲染这个动态生成的包围盒模型,就可以让 Stencil Buffer 记录画布上各个像素位置是否位于阴影中,如图 7-16 所示。

图 7-16 中,场景中有一个点光源 L 位于左上角,一个地板(Floor)上方漂浮着一个立方体。光源 L 照射到 Cube 和 Floor 上,Cube 投射出自己的阴影,这阴影由包围盒描述出来。图中 ABCD 都代表 Floor 上的一点。A 点位于包围盒内部,包围盒的背面的深度测试会失败,所以此处的模板缓存值加 1;包围盒正面的深度测试会成功,所以对模板缓存无影响。B 点比包围盒更靠近 Camera,因此此位置上的包围盒的正面、背面的深度测试都会失败,所以此处的模板缓存值加 1 又减 1,保持为 0。C 点比包围盒更远离 Camera,包围盒的正面、背面的深度测试都会成功,所以此处的模板缓存值保持不变,即为 0。D 点与包围盒的任何一部分都没有交集,因此不受包围盒影响,模板缓存在此位置的值保持不变,即为 0。

包围盒本身是一个模型,但并不存在于原本的场景中,所以在第二次渲染过程中要通过下述代码来避免将其渲染到最终的场景中:

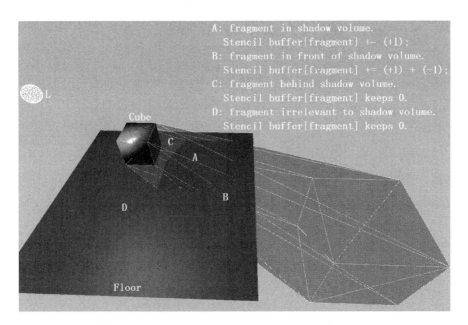

图 7-16　Shadow Volume 判断阴影

```
glDepthMask(false); // Disable writing to Depth Buffer.
glColorMask(false, false, false, false); // Disable writing to Color Buffer.
```

这样就保证了包围盒不改变深度缓存,也不会写入颜色缓存。同时,其"包围阴影"的功能仍然能够正常进行。

为了保证包围盒的正面背面都被渲染,需要禁用面剔除功能:

```
glDisable(GL_CULL_FACE); // Disable culling face.
```

第三遍渲染前,重新设置模板测试的函数和操作:

```
// Draw only if the corresponding stencil value is zero.
glStencilFunc(GL_EQUAL, 0x0, 0xFF);
// prevent updating to theStencil Buffer.
glStencilOp(GL_KEEP, GL_KEEP, GL_KEEP);
```

此时设置模板测试仅允许模板缓存值为 0 的位置通过。也就是说,只有在第二次渲染时位于包围盒外部的模型才会被渲染(即施加光照计算)。此时已经无需也不应该修改模板缓存的值,所以设置为,在任何情形下都让模板缓存的值保持不变。

这时,只需按通常的方式用光照模型渲染场景,即可产生即有光照又有阴影的最终效果。

7.3 模板缓存的初始化

模板缓存、深度缓存和颜色缓存都是 Framebuffer 对象中的成员。如果需要在默认的 Framebuffer 对象中使用模板缓存，必须在创建 Render Context 时就指定要创建的模板缓存。在 Windows 下为默认 Framebuffer 对象指定创建模板缓存的伪代码如下：

```
bool UpdateContextVersion(ContextGenerationParams parameters)
{
    int major, minor;
    GetHighestVersion(out major, out minor);
    if ((major > 2) || (major == 2 && minor > 1)) // for OpenGL version no less than 2.1.
    {
        var attribList = new int[]
        {
            WGL_SUPPORT_OPENGL_ARB, GL_TRUE,
            WGL_DRAW_TO_WINDOW_ARB, GL_TRUE,
            WGL_DOUBLE_BUFFER_ARB, GL_TRUE,
            // something else...
            WGL_DEPTH_BITS_ARB,       parameters.DepthBits,
            WGL_STENCIL_BITS_ARB,     parameters.StencilBits,
                                              // init with Stencil Buffer
            0, //End
        };

        IntPtr dc = this.DeviceContextHandle;
        // Match an appropriate pixel format
        var pixelFormat = new int[1];
        var numFormats = new uint[1];
        if (false == wglChoosePixelFormatARB(dc, attribList, null, 1, pixelFormat,
                numFormats))
        { return false; }

        // ...
    }
}
```

创建 Render Context 是一个坎坷的过程。如果在创建 Render Context 时没有指定创建模板缓存，那么之后对模板缓存的任何操作都是无效的。

7.4 多光源下的阴影

无论 Shadow Mapping 还是 Shadow Volume 都可以简单地应用到有多个光源的场景中。类似于多光源下的光照模型 Blinn-Phong,只需分别对每个光源执行一遍 Shadow Mapping 或 Shadow Volume,并且用混合功能将各个光源的照射效果叠加即可。其伪代码如下:

```
void MultipleLights(Scene scene, ..)
{
    // render ambient color...

    foreach (var light in scene.Lights)
    {
        // preparation...

        glEnable(GL_BLEND); // enable blending.
        glBlend(GL_ONE, GL_ONE); // add lighting info to previous lights.

        // light the scene up with specified light.
        RenderUnderLight(scene, light, ..);

        glDisable(GL_BLEND); // disable blending.
    }
}
```

图 7-17 展示了同时用红绿蓝三色光源照射模型的场景。

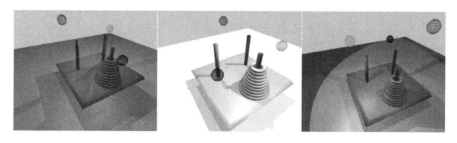

图 7-17　多光源照射的光和影　(左)点光源　(中)平行光　(右)聚光灯

图中用三个小球描述了光源的位置。对于平行光,小球描述的是光源的方向。

7.5 总　结

本章介绍了 Shadow Mapping 和 Shadow Volume 这两种渲染阴影的方法。Shadow Mapping 的实现相对简单，但是阴影的分辨率有限；Shadow Volume 的实现相对复杂，对模型的规范性有一定的要求，但是阴影的分辨率很高。如果将 Shadow Mapping 类比作位图，那么 Shadow Volume 可以类比作矢量图。

① Shadow Mapping 的思路是什么？

两遍渲染：首先从光源位置渲染场景，得到深度缓存；然后依据深度缓存判断 Fragment 是否位于阴影中。

② Shadow Volume 的思路是什么？

三遍渲染：首先渲染场景到深度缓存；然后动态生成阴影包围盒，并设置模版缓存；最后依据模版缓存的状态判断 Fragment 是否位于阴影中。

③ 多光源的阴影如何实现？

依次对每个光源运用 Shadow Mapping 或 Shadow Volume 算法。

④ Shadow Volume 最可能失败的原因是什么？

创建 OpenGL Render Context 时没有指定创建模版缓存。

7.6 问　题

带着问题实践是学习 OpenGL 的最快方式。这里给读者提出几个问题，作抛砖引玉之用。

1. 请在本书网络资料中的 Demos\LogicOperation 项目中尝试使用各种类型的逻辑操作，观察各自的效果。注意，需要将鼠标移动到 Cube 模型上才能看到逻辑操作的效果。

2. 任意选择一个示例项目，或者新建一个项目，尝试使用剪切测试（Scissor Test），然后观察效果。思考剪切测试能够帮助实现什么功能？

3. 在关于模版测试的示例项目 Demos\StencilTest 中，Cube 的包围框的宽度随 Cube 的远离而逐渐减小。请尝试使用 Shader 来保证包围框的宽度保持不变。

第 8 章

拾 取

学完本章之后,读者将能:
- 使用(Color-Coded-Picking)技术拾取(Picking)场景中的模型;
- 通过拖拽的方式修改模型中单个顶点的位置。

8.1 基 础

在 Legacy OpenGL 中,可以用 GL_SELECT 模式实现选择不同的模型,但是不能选择组成模型的单个图元(PRIMITIVE)。本书几乎全部围绕 Modern OpenGL (即使用 Shader 的 OpenGL)来讲述。因此本章介绍一种在 Modern OpenGL 中实现拾取功能的算法,即基于颜色编码的拾取方法。

8.1.1 渲染模式

OpenGL 支持多种渲染模式(Draw Mode),用于渲染不同类型的图元。本书涉及的 Draw Mode 如图 8-1 所示。

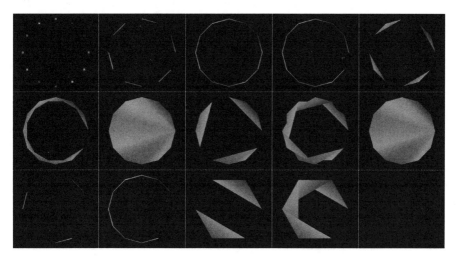

图 8-1 本书涉及的 Draw Mode 一览

如图 8-1 所示，12 个顶点按逆时针方向，从最右侧开始，排列成圆形的轨迹，分别以不同的 Draw Mode 渲染，得到了各自的图形。按从左到右、从上到下的顺序，这些 Draw Mode 的类型依次为：

- GL_POINTS、GL_LINES、GL_LINE_LOOP、GL_LINE_STRIP、GL_TRIANGLES；
- GL_TRIANGLE_STRIP、GL_TRIANGLE_FAN、GL_QUADS、GL_QUAD_STRIP、GL_POLYGON；
- GL_LINES_ADJACENCY、GL_LINE_STRIP_ADJACENCY、GL_TRIANGLES_ADJACENCY、GL_TRIANGLE_STIRP_ADJACENCY、GL_PATCH。

这些 Draw Mode 渲染的图元类型依次为：

- POINT、LINE、LINE、LINE、TRIANGLE；
- TRIANGLE、TRIANGLE、QUAD、QUAD、GL_POLYGON；
- LINE、LINE、TRIANGLE、TRIANGLE。

其中 GL_PATCH 模式并没有显示出任何图形，因为它不是常规的渲染模式，它需要与更新的 Pipeline 的内容（细分着色器）结合才能发挥作用。本章不涉及这些内容。

其中 GL_LINES_ADJACENCY、GL_LINE_STRIP_ADJACENCY、GL_TRIANGLES_ADJACENCY 和 GL_TRIANGLE_STIRP_ADJACENCY 是附带了邻接顶点的 PRIMITIVE 类型，与对应的不含_ADJACENCY 的 PRIMITIVE 类型相似。在介绍 Shadow Volume 算法时使用了 GL_TRIANGLES_ADJACENCY 模式，但是本章不涉及这些内容。

图 8-1 中，第二行第一个图由 GL_TRIANGLE_STRIP 模式渲染的三角形带，

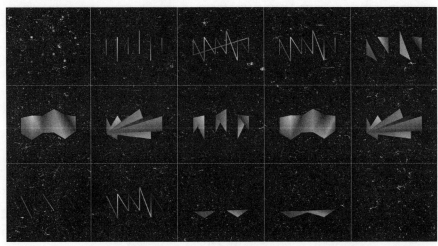

图 8-2　按从上到下、从左到右的顺序安排 12 个顶点

由于顶点的排列位置关系,相邻的三角形产生了一部分的覆盖。第二行第四个由 GL_QUAD_STRIP 模式渲染的四边形带,也产生了交叉的四边形。如果按从上到下、从左到右的顺序安排 12 个顶点,就会正常显示三角形带和四边形带。

图 8-2 中,12 个顶点按从上到下、从左到右的顺序安排,分别以不同的 Draw Mode 渲染,得到了各自的图形。按从左到右、从上到下的顺序,这些 Draw Mode 渲染的 PRIMITIVE 类型与图 8-1 中的示例相同。注意第二行第一个是三角形带,第二行第四个是四边形带,两者的总体形状相同,但是颜色有所不同。第二行第三个由 GL_QUADS 模式渲染的四边形,由于顶点的排列位置关系,产生了交叉的四边形。

8.1.2 渲染命令

OpenGL 还支持多种渲染命令(由接口 IDrawCommand 抽象表示),为渲染 PRIMITIVE 提供了灵活的选项。本书涉及到的渲染命令列表如下:

```
void glDrawArrays(uint mode, int first, int count);
void glDrawArraysInstanced(uint mode, int first, int count, int primcount);
void glDrawElements(uint mode, int count, uint type, IntPtr indices);
void glDrawElementsInstanced(uint mode, int count, uint type, IntPtr indices, int primcount);
void glDrawElementsBaseVertex(uint mode, int count, uint type, IntPtr indices, int basevertex);
void glDrawElementsInstancedBaseVertex(uint mode, int count, uint type, IntPtr indices, int primcount, int basevertex);
```

渲染命令 glDrawArrays(..) 指示 OpenGL 以何种渲染模式从某一位置开始 (first)渲染多少个顶点,它由 DrawArraysCmd 类型封装。

渲染命令 glDrawElements(..) 指示 OpenGL 以何种渲染模式从某一位置开始 (indices)渲染多少个顶点,并且指明 IndexBuffer 的元素类型(type)是 uint、ushort 或 byte,它由 DrawElementsCmd 类型封装。IndexBuffer 封装了 GL_ELEMENT_ARRAY_BUFFER 类型的索引缓存。

其他渲染命令都是这两种渲染命令的变种,通过指定更丰富的参数可以方便实现某些特殊的渲染需求。从面向对象的角度可以视作重载的函数。

8.1.3 认识 gl_VertexID

在 Vertex Shader 中,有一个内置的系统变量 int gl_VertexID,它表示的是当前 Vertex Shader 处理的顶点在 VertexBuffer 中的位置(索引值)。在描述 Pipeline 的伪代码中,有下述一段:

```
...
// execute vertex shader.
vec4[] gl_Positions = new vec4[positions.Length];
for (inti = 0; i < positions.Length; i++) // for each element in VertexBuffer:
{
    var vs = new VertexShader();
    vs.inPosition = positions[i];
    vs.main(); // execute the vertex shader we provided.
    gl_Positions[i] = vs.gl_Position;
}
...
```

其中 for 循环里的变量 i 就是 gl_VertexID 的值。加入 gl_VertexID 后,上述片段可以更新如下:

```
...
// execute vertex shader.
vec4[] gl_Positions = new vec4[positions.Length];
for (inti = 0; i < positions.Length; i++) // for each element in VertexBuffer:
{
    var vs = new VertexShader();
    vs.gl_VertexID = i; // execution order of this vertex shader.
    vs.inPosition = positions[i];
    vs.main(); // execute the vertex shader we provided.
    gl_Positions[i] = vs.gl_Position;
}
...
```

虽然在由显卡驱动实现的 OpenGL 里,此处的 for 循环是并行执行的,但是不妨碍让每次执行的 Vertex Shader 知晓自己处理的是 VertexBuffer 中的第几个顶点。

下面通过一个实验来进一步认识 gl_VertexID 的作用。此实验的关键点是以 gl_VertexID 的值为依据,为顶点着色。创建一个新的模型,它由 10×10×10 个互不相干的顶点组成,这些顶点均匀地分布在以原点为中心的 Model Space 里。顶点的颜色由 2 种渲染方法分别指定。一种是在程序中随机生成一些颜色,另一种则根据 gl_VertexID 的值设置颜色。将 gl_VertexID 转换为颜色值的代码如下:

```
// vertex shader that gets color value according to gl_VertexID.
#version 150 core

in vec3 inPosition;
uniform mat4 mvpMatrix;
out vec4 passColor;
```

```
void main() {
    // transform vertex' position from model space to clip space.
    gl_Position = mvpMatrix * vec4(inPosition, 1.0);

    // gets color value according to gl_VertexID.
    int index = gl_VertexID;
    passColor = vec4(
        float(index & 0xFF) / 255.0,         // red
        float((index >> 8) & 0xFF) / 255.0,  // green
        float((index >> 16) & 0xFF) / 255.0, // blue
        float((index >> 24) & 0xFF) / 255.0); // alpha
}
```

这 2 种渲染方法的结果如图 8-3 所示。

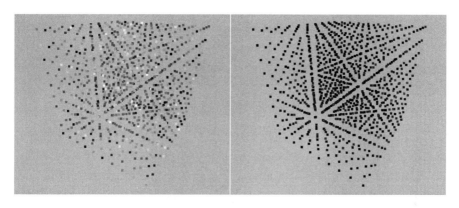

图 8-3　(左)随机颜色　(右)根据 gl_VertexID 上色

由于 gl_VertexID 是 int 类型，有 4 个字节，恰好可以让一个字节对应描述颜色值的 RGBA 的一个分量。由于 gl_VertexID 的取值范围是 int 类型的所有非负数值，所以本模型里十分有限的顶点在第二种渲染方法中呈现的都接近红色。为了便于观察和操作，这里借助 Shader 里的内置变量 gl_PointSize 来增大点的面积，使之呈现为一个明显的正方形。

此示例也引出在 OpenGL 中常用的一个调试技术，即借助输出的颜色来分析 Pipeline 的执行过程。由于 OpenGL 是由显卡驱动执行的，不能像常规程序那样进行单步调试，这种借助输出的分析技术就成为了一个很重要的手段。请读者打开本书网络资料上的 c08d00_gl_VertexID 项目，查看此实验项目的完整代码。

8.1.4　认识 flat

OpenGL Shader 中的关键字 flat 的作用类似函数 glShadeMode(GL_FLAT)，

它使得一个 PRIMITIVE 在渲染时,其颜色由构成自身的最后一个顶点的颜色决定(GL_POLYGON 是例外,由第一个顶点决定),且整个 PRIMITIVE 全部都是这个颜色。此时,线性插值功能就不再对颜色进行插值处理了。例如,对于图 8-1 所示的各种 PRIMITIVE,如果使用 flat 关键字,那么显示出的颜色将如图 8-4 所示。

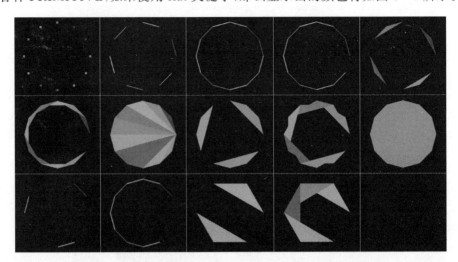

图 8-4　用 flat 关键字渲染的 PRIMITIVE

注意,在不使用 flat 时,GL_TRIANGLE_FAN 和 GL_POLYGON 显示出了同样的着色结果。但在使用 flat 时,GL_TRIANGLE_FAN 的每个 PRIMITIVE 都是三角形,其颜色分别由构成自身的最后一个顶点的颜色决定;而 GL_POLYGON 渲染的是个单一的 PRIMITIVE(即一个多边形),其颜色由第一个顶点的颜色决定。

使用 flat 关键字十分简单,只需在需要的自定义"in\out"属性变量前加上 flat 即可。例如在上一节中将 gl_VertexID 转换为颜色值的代码,只需做如下改动:

```
// vertex shader that gets color value according to gl_VertexID.
#version 150 core

in vec3 inPosition;
uniform mat4 mvpMatrix;

flat out vec4 passColor; // use flat keyword.

void main() {
    ...
}
```

只需在 out vec4 passColor; 变量前添加 flat 关键字即可。注意,相应地要在 Fragment Shader 的对应变量中也添加 flat 关键字,如下所示:

```
//fragment shader that gets color value according to gl_VertexID.
#version 150 core

flat in vec4 passColor;// use flat keyword.

out vec4 outColor;

voidmain() {
    outColor = passColor;
}
```

读者可在本书网络资料上打开 c08d04_DrawModes 项目,对比观察使用 flat 和不使用 flat 的渲染效果。

8.1.5 认识 glReadPixels(..)函数

```
// read pixels data in specified format and type from location(x, y) and region(width, height).
    glReadPixels(int x, int y, int width, int height, uint format, uint type, IntPtr pixels);
```

使用 OpenGL 函数 glReadPixels(..) 可以获取当前 Framebuffer 上指定位置或区域的颜色信息。这个颜色信息,是可以包含不在画布上直接展示的 alpha 分量的。一般的,在需要读取单个像素所在位置上的颜色值时,可以使用如下代码:

```
// C# code.
struct Pixel
{
    public byte r;
    public byte g;
    public byte b;
    public byte a;
}

// Reads pixel's color in Color Buffer at specified position(x, y) via glReadPixels(..)
Pixel ReadPixel(int x, int y)
{
    var array = new Pixel[1];
    GCHandle pinned = GCHandle.Alloc(array, GCHandleType.Pinned);// array can't move.
    IntPtr header = pinned.AddrOfPinnedObject(); // unmanged address of array.
    const int width = 1, height = 1;
    const uint format = GL_RGBA, type = GL_UNSIGNED_BYTE;
```

```
    // get color in Color Buffer of current framebuffer.
    glReadPixels(x, y, width, height, format, type, header);
    pinned.Free(); // array can move now. GC can work on it.

    return array[0]; // result is in the only element of array.
}
```

C#代码是托管的,C#创建的数组 Pixel[]随时可能被GC移动到内存里的其他位置。而 glReadPixels(..)需要向一个固定位置写入它得到的结果,因此需要用 GCHandle 提供的方法将数组 Pixel[]临时固定。同时,GCHandle 也提供了数组 Pixel[]的指针 header。此指针不在托管范围内,它类似于 C\C++中的指针,指向的是已经固定了的数组 Pixel[]在内存中的地址,因此可以传入 glReadPixels(..)。

使用 glReadPixels(..)还可以直接获取整个画布的内容,进而可以保存为一个图片文件。也就是说,可以实现截屏的功能,代码如下:

```
// C# code.
public static void Save2Picture(int x, int y, int width, int height, string filename)
{
    var format = System.Drawing.Imaging.PixelFormat.Format32bppArgb;
    var bitmap = new Bitmap(width, height, format);
    var lockMode = System.Drawing.Imaging.ImageLockMode.WriteOnly;
    var rect = new Rectangle(0, 0, bitmap.Width, bitmap.Height);
    var bmpData = bitmap.LockBits(rect, lockMode, format); // lock position in memory.

    glReadPixels(x, y, width, height, GL_BGRA, GL_UNSIGNED_BYTE, bmpData.Scan0);
    bitmap.UnlockBits(bmpData); // unlock.
    bitmap.RotateFlip(RotateFlipType.Rotate180FlipX); // so that bitmap fits in windows space.

    bitmap.Save(filename);
}
```

8.1.6 Color-Coded-Picking 算法基础

基于上述基本情况,Color-Coded-Picking 算法以 gl_VertexID 内置变量、flat 关键字和 glReadPixels(..) 函数为基础,实现了在任意 Draw Mode 和任意 IDrawCommand 下拾取任意几何体(Geometry)的功能。这里的几何体包括:点(Point)、线(Line)、三角形(Triangle)、四边形(Quad)和多边形(Polygon)共5种。

注意,图元(PRIMITIVE)与几何体(Geometry)是不同的概念。OpenGL 渲染的是 PRIMITIVE,拾取功能拾取到的是 Geometry,在一个 PRIMITIVE 中可能拾取多种 Geometry。例如,可以在 GL_TRIANGLES 模式下渲染的一个 TRIANGLE 里拾取一个三角形,也可以拾取这个 TRIANGLE 的一个顶点或一条边,但是不能在 GL_TRIANGLES 模式下拾取四边形或多边形。

不同的 PRIMITIVE 里能够拾取的 Geometry 类型,如表 8-1 所列。

表 8-1 PRIMITIVE 里能够拾取的 Geometry

PRIMITIVE \ Geometry	Point	Line	Triangle	Quad	Polygon
POINT	√				
LINE	√	√			
TRIANGLE	√	√	√		
QUAD	√	√		√	
POLYGON	√	√			√

由于 Draw Mode 和 IDrawCommand 的种类较多,Color-Coded-Picking 算法需要分别针对每种情况编写相应的识别函数,这是导致此算法比较复杂的原因。运用面向对象的编程方法,是降低复杂程度的有效途径。

用 GL_POINTS 模式,可以渲染出一个个的点,这是最简单的渲染模式。本节以最简单的 GL_POINTS 为例,说明 Color-Coded-Picking 的基本原理。通过 glReadPixels(..)函数,可以获得鼠标所在位置的颜色值 Pixel(r, g, b, a)。如果这个颜色表示的是鼠标指向的顶点在 VertexBuffer 中的索引,那么就实现了拾取功能。

用 8.1.3 小节中的方式,根据 Vertex Shader 中的内置变量 gl_VertexID 来设置颜色,颜色就隐含了此顶点的索引值。根据此颜色,完全可以实施一个简单的逆向工程,反推得到 gl_VertexID 的值,代码如下:

```
// C# code.
// translate Pixel(r, g, b, a) color to gl_VertexID
int ToVertexId(byte r, byte g, byte b, byte a)
{
    int shiftedR = r;
    int shiftedG = g << 8;
```

```
        int shiftedB = b << 16;
        int shiftedA = a << 24;
        int gl_VertexID = shiftedR + shiftedG + shiftedB + shiftedA;

        return gl_VertexID;
}
```

纵观全程,获取鼠标所在位置顶点信息的过程如图 8-5 所示。

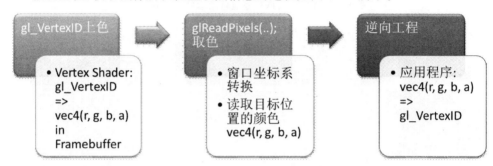

图 8-5　Color-Coded-Picking 原理

注意,OpenGL 对窗口的设定,是以左下角为原点(0,0),以向右为 X 轴正方向,以向上为 Y 轴正方向。而 WinForm 是以左上角为原点(0,0),以向右为 X 轴正方向,以向下为 Y 轴正方向,如图 8-6 所示。

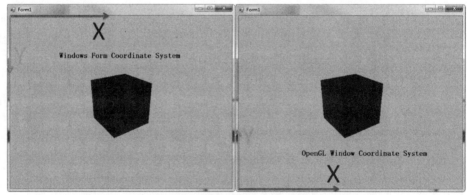

图 8-6　(左)Windows 窗口坐标系　(右)OpenGL 窗口坐标系

因此在计算鼠标位置时要进行转换,代码如下:

```
void glCanvas1_MouseMove(object sender, MouseEventArgs e)
{
    int x = e.X;
    // transform from window coordinate system to OpenGL window coordinate system.
    int y = this.winGLCanvas1.Height - e.Y - 1;
    // ...
}
```

如果暂时忽略场景管理等问题，现在已经实现了拾取单个顶点的功能。读者可在本书网络资料的 c08d01_PickPoint 项目中查看具体效果，如图 8-7 所示。

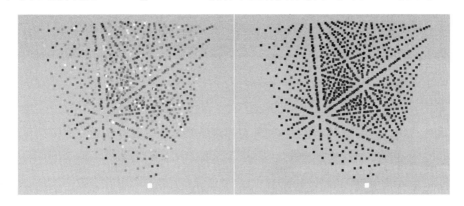

图 8-7　用鼠标拾取顶点　（左）随机颜色　（右）顺序颜色

如图 8-7 所示，鼠标移动到点集的最下方的点上，此点显示为白色，并且增大面积以突出其被选中的状态。此项目的窗口会展示此点的具体信息，如下：

```
CSharpGL - picked: Stage Vertex ID: 9
From: [1]: [PointsNode]
Geometry Type: Point
Positions in Model Space:
[9]: -4.5000, -4.5000, 4.5000
```

这些信息说明，拾取到的顶点的索引值为 9，它在 Model Space 里的位置为 vec3(-4.5, -4.5, 4.5)。如果读者检查一遍 PointsNode 的定义，会发现客户端布置的 positions[9] 的值正是 vec3(-4.5, -4.5, 4.5)，这验证了拾取操作的正确性。在应用程序端实现拾取功能的关键代码如下：

```
private Picking pickingAction; // do the Color-Coded-Picking.
private PointsNode pickedNode; // node in which a point was picked.

void winGLCanvas1_MouseMove(object sender, MouseEventArgs e)
{
    int x = e.X;
    int y = this.winGLCanvas1.Height - e.Y - 1;
    // try to pick something at specified location.
    PickedGeometry pickedGeometry = this.pickingAction.Pick(x, y, ..);

    if (pickedGeometry != null) // if something's picked.
    {
        // node that contains the picked point.
```

```
            var node = pickedGeometry.FromObject as PointsNode;
            if (node != null)
            {
                // tell the node about the vertex id of the picked point.
                node.HighlightIndex = (int)pickedGeometry.StageVertexId;
                this.pickedNode = node;
            }
        }
        else
        {
            if (this.pickedNode != null)
            {
                // tell the node that nothing is picked.
                this.pickedNode.HighlightIndex = -1;
            }
        }
    }
```

每当鼠标在画布上移动时，Picking对象就会执行一次拾取操作，如果鼠标所在位置上有顶点，那么就会将其高亮显示，否则就让上次选中的顶点恢复正常显示。实现了拾取功能的Picking对象，在内部完成了如下操作：

```
// Pick geometry at specifiedposition(x, y).
public PickedGeometry Pick(int x, int y, ..)
{
    Scene scene = … // Gets the scene somehow.
    PickedGeometry pickedGeometry = null;
    
    Framebuffer framebuffer = GetPickingFramebuffer(width, height);
    framebuffer.Bind();
    {
        glClearColor(1.0f, 1.0f, 1.0f, 1.0f); // white color means nothing picked.
        glClear(GL_COLOR_BUFFER_BIT | GL_DEPTH_BUFFER_BIT);
        // render the scene using gl_VertexID method.
        this.RenderForPicking(scene.RootNode, ..);
        // Gets the gl_VertexID using glReadPixels(..) and (r,g,b,a) => gl_VertexID.
        uint gl_VertexID = ColorCodedPicking.ReadVertexId(x, y);
        // Gets the primitive's information at specified gl_VertexID.
        pickedGeometry = GetPickedGeometry(gl_VertexID, ..);
    }
    framebuffer.Unbind();
```

```
        return pickedGeometry;
    }
    // Gets the primitive's information at specified gl_VertexID
    private public PickedGeometry GetPickedGeometry(uint vertexId, ..)
    {
        var vertexIds = new uint[] { vertexId, }; // indexes.
        vec3[] positions = GetPositions(vertexId); // positions.
        var pickedGeometry = new PickedGeometry(GeometryType.Point, positions, vertex-
                            Ids, ..);
        return pickedGeometry;
    }
    // Gets the positions of the pickedprimitive
    private vec3[] GetPositions (params uint[] positionIndexes)
    {
        VertexBuffer positionBuffer = ... // gets position buffer somehow
        var positions = new vec3[positionIndexes.Length];
        //mapping vertex buffer to app memory address so that we can access to it.
        IntPtr pointer = positionBuffer.MapBuffer(MapBufferAccess.ReadOnly);
        unsafe
        {
            var array = (vec3 *)pointer.ToPointer();
                            // we got the array representing the vertex buffer
            for (var i = 0; i < positionIndexes.Length; i++)
            {
                // fill in the positions(for the GL_POINTS, actually only 1 position).
                positions[i] = array[ positionIndexes[i] ];
            }
        }
        positionBuffer.UnmapBuffer();
                    // unmapping vertex buffer so that OpenGL can work with it

        return positions;
    }
```

即使仅仅针对 GL_POINTS 这种最简单的渲染模式,其拾取过程也是比较复杂的。

为了实施 Color-Coded-Picking,首先需要用 gl_VertexID 方式渲染整个场景。也就是说,每个支持拾取功能的 Node,都要单独实现一个用 gl_VertexID 着色的 RenderMethod。为了避免影响用户看到的情景,这里需要创建一个自定义的 Framebuffer,在此 Framebuffer 下渲染场景。

根据 Color-Coded-Picking 算法,vec4(0 / 255.0, 0 / 255.0, 0 / 255.0, 0 /

255.0)代表VertexBuffer中第一个顶点,vec4(255 / 255.0, 255 / 255.0, 255 / 255.0, 255 / 255.0)代表可能的最后一个顶点,而vec4(255 / 255.0, 255 / 255.0, 255 / 255.0, 255 / 255.0)在OpenGL Shader中描述的就是白色。一般的,VertexBuffer不可能达到 $256 \times 256 \times 256 \times 256 = 2^{32}$ 这样的长度,因此用白色代表没有选中任何顶点是最自然的。

获取鼠标所在位置的颜色值,进而获取gl_VertexID值是简单的。此时,gl_VertexID是选中的顶点在VertexBuffer中的索引,如果单纯告知用户这个索引,是没有直接意义的。因此,这里继续挖掘,找到VertexBuffer中的数据,即此顶点在Model Space里的位置。为了从VertexBuffer中读取数据,使用了 glMapBuffer(..) 和 glUnmapBuffer() 函数。通过glMapBuffer(..),可以获取代表VertexBuffer的数组的指针,并且像使用普通的内存中的数组一样使用它。使用完毕后,用glUnmapBuffer()打扫战场,让OpenGL恢复VertexBuffer的状态。

在编写应用程序时,还会遇到一个细节问题:在有多个Node的场景中拾取时,如何区分多个Node?渲染任何一个Node时,都要启动一次Pipeline,其Vertex Shader里的gl_VertexID都是从零开始计数的。试想,如果对每个Node,都按照上文的算法执行gl_VertexID渲染方法,那么,所有Node的第N个顶点都会是相同的颜色。如何区分这个Node的第N个顶点和那个Node的第N个顶点呢?

答案很简单,只需在渲染每个Node时,为其指定一个提前量,告诉它在它之前已经渲染了多少个顶点,让它接着渲染即可。据此修改后的gl_VertexID渲染方法的Vertex Shader如下:

```glsl
// vertex shader that gets color value according to gl_VertexID.
#version 150 core

in vec3 inPosition;
uniform mat4 mvpMatrix;
uniform int pickingBaseId;// next id to start with.

out vec4 passColor;

void main() {
    // transform vertex' position from model space to clip space.
    gl_Position = mvpMatrix * vec4(inPosition, 1.0);

    // gets color value according to gl_VertexID.
    int index = pickingBaseId + gl_VertexID;
    passColor = vec4(
        float(index & 0xFF) / 255.0,
        float((index >> 8) & 0xFF) / 255.0,
        float((index >> 16) & 0xFF) / 255.0,
        float((index >> 24) & 0xFF) / 255.0);
}
```

只需新增一个 uniform int pickingBaseId; 变量,即可保证场景中所有 Node 的所有顶点的颜色都各不相同,进而确保能够区分出选中的顶点属于哪个 Node。相应地,GetPickedGeometry(..)也要做出修改,代码如下:

```
// Gets picked geometry in the scene.
private PickedGeometry GetPickedGeometry(uint stageVertexId, SceneNodeBase node, ..)
{
    PickedGeometry pickedGeometry = null;
    if (node != null)
    {
        var pickable = node as IPickable;
        if (pickable != null) // if this node supports Color-Coded-Picking.
        {
            uint baseId = pickable.PickingBaseId; // first index of this node.
            // if the picked geometry is in this node.
            if (baseId <= stageVertexId && stageVertexId < baseId + pickable.VertexCount)
            {
                pickedGeometry = pickable.GetPickedGeometry(stageVertexId, baseId, ..);
            }
        }

        if (pickedGeometry == null) // if the picked geometry is not in the node.
        {
            foreach (var item in node.Children) // check if it's in node's children.
            {
                pickedGeometry = GetPickedGeometry(stageVertexId, item, ..);
                if (pickedGeometry != null) { break; } // we found it!
            }
        }
    }

    return pickedGeometry;
}
```

基于修改后的 gl_VertexID 渲染方法,可以快速地筛选出选中的顶点所在的 Node。然后在此 Node 中详细查找选中的顶点信息的过程,就是在重复上文的单个 Node 的情形。总之,对于有多个 Node 的场景,可以将其视作一个虚拟的单独的 Node,这样就可以继续用 Color-Coded-Picking 算法实现拾取功能。

8.2 在 DrawArraysCmd 命令下的拾取

```
// drawc ount vertexes in mode from the first vertex in VertexBuffer.
void glDrawArrays(uint mode, int first, int count);
```

DrawArraysCmd 对象封装了 glDrawArrays(..)命令,这是最简单的渲染命令,它指示 OpenGL 按照顶点在 VertexBuffer 中的顺序渲染。在以 gl_VertexID 渲染方法渲染场景后,顶点按照 VertexBuffer 中的索引顺序构成各个 PRIMITIVE。也就是说,PRIMITIVE 的各个顶点在 VertexBuffer 中的位置是连续的。

8.2.1 拾取 GL_POINTS 模式下的 Geometry

GL_POINTS 模式下渲染的 PRIMITIVE 是若干 POINT,能够拾取的 Geometry 只有 Point。在 8.1.6 小节已经介绍了拾取 GL_POINTS 模式下的 Point 的方法,这里不再赘述。

8.2.2 拾取 GL_LINES 模式下的 Geometry

GL_LINES 模式下渲染的 PRIMITIVE 是若干 LINE,能够拾取的 Geometry 是 Line 和 Point。如果想拾取 Line,应以 GL_LINES 模式执行 gl_VertexID 渲染方法。由于采用了 flat 关键字,此时无论鼠标位于 LINE 的哪一点上,反推计算得到的 gl_VertexID 都是要拾取的 Line 的最后一个(即第二个)顶点的索引。那么要拾取的 Line 的 2 个顶点的索引分别是[gl_VertexID-1]和[gl_VertexID],如图 8-8 所示。

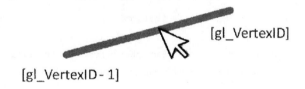

图 8-8　GL_LINES 模式下拾取 Line

如果想拾取 Point,应以 GL_POINTS 模式执行 gl_VertexID 渲染方法。此时,就回到了最简单的拾取情形(即在 GL_POINTS 模式下拾取 Point)。处理方法参见 8.1.6 小节。

8.2.3 拾取 GL_LINE_LOOP 模式下的 Geometry

GL_LINE_LOOP 模式下渲染的 PRIMITIVE 是若干 LINE,能够拾取的 Geometry 是 Line 和 Point。如果想拾取 Line,应以 GL_LINE_LOOP 模式执行 gl_VertexID 渲染方法。由于采用了 flat 关键字,此时无论鼠标位于 LINE 的哪一点上,反推计算得到的 gl_VertexID 都是要拾取的 Line 的最后一个(即第二个)顶点的索引。那么要拾取的 Line 的 2 个顶点的索引分别是[gl_VertexID - 1]和[gl_VertexID],如图 8-9 所示。

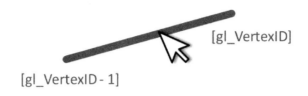

图 8-9　GL_LINE_LOOP 模式下拾取 Line

但是如果反推得到的 gl_VertexID 为 0,那就意味着拾取到的是由模型的最后一个顶点和第一个顶点组成的 Line,此时要拾取的 Line 的 2 个顶点的索引分别是[VertexCount - 1]和[0],如图 8-10 所示。

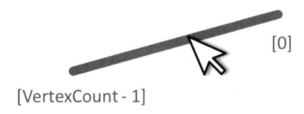

图 8-10　GL_LINE_LOOP 模式下拾取最后一个 Line

如果想拾取 Point,应以 GL_POINTS 模式执行 gl_VertexID 渲染方法。此时,就回到了最简单的拾取情形(即在 GL_POINTS 模式下拾取 Point)。处理方法参见 8.1.6 小节。

8.2.4　拾取 GL_LINE_STRIP 模式下的 Geometry

GL_LINE_STRIP 模式下渲染的 PRIMITIVE 是若干 LINE,能够拾取的 Geometry 是 Line 和 Point。此时与拾取 GL_LINES 模式下的 Geometry 情形类似,处理方法完全相同,不再赘述。

8.2.5　拾取 GL_TRIANGLES 模式下的 Geometry

GL_TRIANGLES 模式下渲染的 PRIMITIVE 是若干 TRIANGLE,能够拾取的 Geometry 是 Triangle、Line 和 Point。

如果想拾取 Triangle,应以 GL_TRIANGLES 模式执行 gl_VertexID 渲染方法。由于采用了 flat 关键字,此时无论鼠标位于 TRIANGLE 的哪一点上,反推计算得到的 gl_VertexID 都是要拾取的 Triangle 的最后一个(即第三个)顶点的索引。那么要拾取的 Triangle 的 3 个顶点的索引分别是[gl_VertexID - 2]、[gl_VertexID - 1]和

[gl_VertexID],如图 8-11 所示。

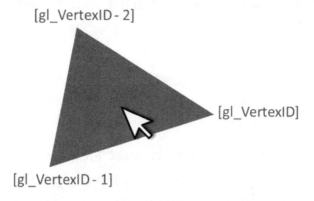

图 8-11　GL_TRIANGLES 模式下拾取 Triangle

如果想拾取 Line,情况就有些复杂了。应启用 glPolygonMode(GL_LINE),然后以 GL_TRIANGLES 模式执行 gl_VertexID 渲染方法。此时,只有 TRIANGLE 的各个边会被渲染。由于采用了 flat 关键字,当鼠标位于 TRIANGLE 的某一条边上时,反推计算得到的 gl_VertexID 就是 TRIANGLE 的最后一个(即第三个)顶点的索引。那么此 TRIANGLE 的 3 个顶点的索引分别是[gl_VertexID - 2][gl_VertexID - 1]和[gl_VertexID]。为了判定拾取的 Line 是此 TRIANGLE 里的哪一个,可以设计一个临时的渲染命令,使之以 GL_LINES 模式渲染这三个顶点组成的 3 个 LINE,如图 8-12 所示。

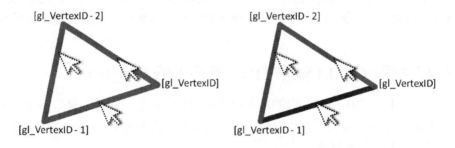

图 8-12　(左)拾取到线框状态的 TRIANGLE　(右)拾取以 GL_LINES 模式渲染的 3 个 LINE

此时,3 个 LINE 分别由{gl_VertexID, gl_VertexID - 2}{gl_VertexID - 2, gl_VertexID - 1}和{gl_VertexID - 1, gl_VertexID}组成。这就转化为 8.2.2 小节中的问题。

如果想拾取 Point,应以 GL_POINTS 模式执行 gl_VertexID 渲染方法。此时,就回到了最简单的拾取情形(即在 GL_POINTS 模式下拾取 Point)。处理方法参见 8.1.6 小节。

8.2.6 拾取 GL_TRIANGLE_STRIP 模式下的 Geometry

GL_TRIANGLE_STRIP 模式下渲染的 PRIMITIVE 是若干 TRIANGLE,能够拾取的 Geometry 是 Triangle、Line 和 Point。此时与拾取 GL_TRIANGLES 模式下的 Geometry 情形类似,处理方法完全相同,不再赘述。

8.2.7 拾取 GL_TRIANGLE_FAN 模式下的 Geometry

GL_TRIANGLE_FAN 模式下渲染的 PRIMITIVE 是若干 TRIANGLE,能够拾取的 Geometry 是 Triangle、Line 和 Point。如果想拾取 Triangle,应以 GL_TRIANGLE_FAN 模式执行 gl_VertexID 渲染方法。由于采用了 flat 关键字,此时无论鼠标位于 TRIANGLE 的哪一点上,反推计算得到的 gl_VertexID 都是要拾取的 Triangle 的最后一个(即第三个)顶点的索引。那么要拾取的 Triangle 的 3 个顶点的索引分别是[0][gl_VertexID-1]和[gl_VertexID],如图 8-13 所示。

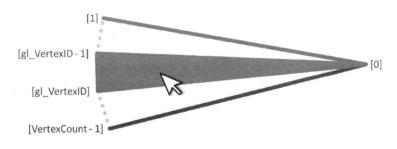

图 8-13　GL_TRIANGLE_FAN 模式下拾取 Triangle

如果想拾取 Line,情况就有些复杂。应启用 glPolygonMode(GL_LINE),然后以 GL_TRIANGLE_FAN 模式执行 gl_VertexID 渲染方法。此时,只有 TRIANGLE 的各个边会被渲染。由于采用了 flat 关键字,当鼠标位于 TRIANGLE 的某一条边上时,反推计算得到的 gl_VertexID 就是 TRIANGLE 的最后一个(即第三个)顶点的索引。那么此 TRIANGLE 的 3 个顶点的索引分别是[0][gl_VertexID-1]和[gl_VertexID]。为了判定拾取的 Line 是此 TRIANGLE 的哪一个,可以设计一个临时的渲染命令,使之以 GL_LINES 模式渲染这三个顶点组成的 3 个 LINE,如图 8-14 所示。

此时,3 个 LINE 分别由{gl_VertexID, 0}{0, gl_VertexID-1}和{gl_VertexID-1, gl_VertexID}组成。这就转化为 8.2.2 小节中的问题。

如果想拾取 Point,应以 GL_POINTS 模式执行 gl_VertexID 渲染方法。此时,就回到了最简单的拾取情形(即在 GL_POINTS 模式下拾取 Point),处理方法参见 8.1.6 小节。

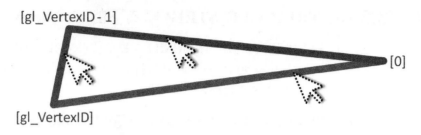

图 8 – 14 GL_TRIANGLE_FAN 模式下拾取 Line

8.2.8 拾取 GL_QUADS 模式下的 Geometry

GL_QUADS 模式下渲染的 PRIMITIVE 是若干 QUAD,能够拾取的 Geometry 是 Quad、Line 和 Point。如果想拾取 Quad,应以 GL_QUADS 模式执行 gl_VertexID 渲染方法。由于采用了 flat 关键字,此时无论鼠标位于 QUAD 的哪一点上,反推计算得到的 gl_VertexID 都是要拾取的 Quad 的最后一个(即第四个)顶点的索引。那么要拾取的 Quad 的 4 个顶点的索引分别是[gl_VertexID – 3][gl_VertexID – 2] [gl_VertexID – 1]和[gl_VertexID],如图 8 – 15 所示。

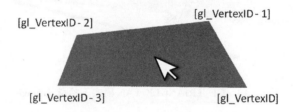

图 8 – 15 GL_QUADS 模式下拾取 Quad

如果想拾取 Line,情况就有些复杂。应启用 glPolygonMode(GL_LINE),然后以 GL_QUADS 模式执行 gl_VertexID 渲染方法。此时,只有 QUAD 的各个边会被渲染。由于采用了 flat 关键字,当鼠标位于 QUAD 的某一条边上时,反推计算得到的 gl_VertexID 就是 QUAD 的最后一个(即第四个)顶点的索引。那么此 QUAD 的 4 个顶点的索引分别是[gl_VertexID – 3][gl_VertexID – 2][gl_VertexID – 1]和 [gl_VertexID]。为了判定拾取的 Line 是此 QUAD 里的哪一个,可以设计一个临时的渲染命令,使之以 GL_LINES 模式渲染这四个顶点组成的 4 个 LINE,如图 8 – 16 所示。

此时,4 个 LINE 分别由{gl_VertexID, gl_VertexID – 3}{gl_VertexID – 3, gl_VertexID – 2}、{gl_VertexID – 2, gl_VertexID – 1}和{gl_VertexID – 1, gl_VertexID}组成。这就转化为 8.2.2 小节中的问题。

如果想拾取 Point,应以 GL_POINTS 模式执行 gl_VertexID 渲染方法。此时,

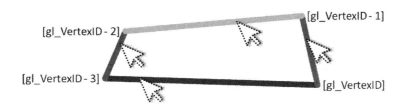

图 8-16 GL_QUADS 模式下拾取 Line

就回到了最简单的拾取情形(即在 GL_POINTS 模式下拾取 Point)。处理方法参见 8.1.6 小节。

8.2.9 拾取 GL_QUAD_STRIP 模式下的 Geometry

GL_QUAD_STRIP 模式下渲染的 PRIMITIVE 是若干 QUAD,能够拾取的 Geometry 是 Quad、Line 和 Point。此时与拾取 GL_QUADS 模式下的 Geometry 情形类似,处理方法完全相同,不再赘述。

8.2.10 拾取 GL_POLYGON 模式下的 Geometry

GL_POLYGON 模式下渲染的 PRIMITIVE 是一个 POLYGON,能够拾取的 Geometry 是 Polygon、Line 和 Point。如果想拾取 Polygon,应以 GL_POLYGON 模式执行 gl_VertexID 渲染方法。由于采用了 flat 关键字,此时无论鼠标位于 POLYGON 的哪一点上,反推计算得到的 gl_VertexID 都是要拾取的 Polygon 的第一个(这与其他模式不同)顶点的索引。那么要拾取的 Polygon 的各个顶点的索引分别是 [gl_VertexID][gl_VertexID + 1][gl_VertexID + 2]……和[gl_VertexID + n - 1],如图 8-17 所示。

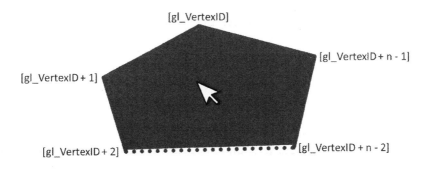

图 8-17 GL_POLYGON 模式下拾取 Polygon

如果想拾取 Line,情况就有些复杂。应启用 glPolygonMode(GL_LINE),然后以 GL_POLYGON 模式执行 gl_VertexID 渲染方法。此时,只有 POLYGON 的各个边会被渲染。由于采用了 flat 关键字,当鼠标位于 POLYGON 的某一条边上时,反推计算得到的 gl_VertexID 就是 POLYGON 的第一个(这与其他模式不同)顶点的索引。那么要拾取的 Polygon 的各个顶点的索引分别是[gl_VertexID][gl_VertexID + 1][gl_VertexID + 2]……和[gl_VertexID + n - 1]。

为了判定拾取的 Line 是此 POLYGON 里的哪一个,可以设计一个临时的渲染命令,使之以 GL_LINE_LOOP 模式渲染这个 POLYGON,如图 8-18 所示。

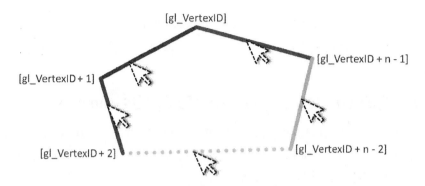

图 8-18 GL_POLYGON 模式下拾取 Line

此时,若干个 LINE 分别由{gl_VertexID, gl_VertexID + 1}{gl_VertexID + 1, gl_VertexID + 2}……{gl_VertexID + n - 2, gl_VertexID + n - 1}和{gl_VertexID + n - 1, gl_VertexID}组成。这就转化为 8.2.3 小节中的问题。

对于 glDrawArrays(..)命令来说,这里的 gl_VertexID 必定是其参数 first 。

如果想拾取 Point,应以 GL_POINTS 模式执行 gl_VertexID 渲染方法。此时,就回到了最简单的拾取情形(即在 GL_POINTS 模式下拾取 Point)。处理方法参见 8.1.6 小节。

8.3 在 DrawElementsCmd 命令下的拾取

```
// Draw count vertex-times in mode from the first vertex indicated by indices
// Also, this method use a GL_ELEMENT_ARRAY_BUFFER to describe the order of vertexes to
    draw.
void glDrawElements(uint mode, int count, uint type, IntPtr indices);
```

DrawElementsCmd 对象封装了 glDrawElements(..)命令,这是借助了 IndexBuffer 的渲染命令。要注意的是,在 Pipeline 的 Vertex Shader 阶段,各个顶点仍旧是按照其在 VertexBuffer 中的顺序依次处理的,每个顶点仍旧只被处理一次。这与 glDrawArrays(..)渲染命令中的处理方式相同。在以 gl_VertexID 渲染方法渲染场景后,顶点按照 IndexBuffer 中描述的索引顺序构成各个 PRIMITIVE。也就是说,一个 PRIMITIVE 的各个顶点在 VertexBuffer 中的位置(即 gl_VertexID 值)未必是连续的,且同一个顶点可能被多个 PRIMITIVE 重复使用。

由于 PRIMITIVE 的各个顶点在 VertexBuffer 中的位置未必是连续的,此时就不能通过 PRIMITIVE 最后一个顶点的 gl_VertexID 直接得到 PRIMITIVE 其他顶点的索引。但是描述此 PRIMITIVE 的全部索引在 IndexBuffer 中是连续存放的,且 gl_VertexID 是最后一个(GL_POLYGON 是例外)。因此,可以通过在 IndexBuffer 中查找 gl_VertexID 的方式来进一步找到 PRIMITIVE 的全部顶点的索引。

DrawElementsCmd 可能启用 PRIMITIVE 的截断功能,即启用下述函数:

```
glEnable(GL_PRIMITIVE_RESTART);
uintindex = uint.MaxValue; // 4294967295
glPrimitiveRestartIndex(index);
```

启用后,在 DrawElementsCmd 使用的 IndexBuffer 中,所有等于 index 的元素都被视为前一个 PRIMITIVE 的结束标志和后一个 PRIMITIVE 的开始标志,即一个截断点。(index 本身不参与构成任何 PRIMITIVE)为记述方便,本章使用 PR 代表 IndexBuffer 中的截断点。一般的,这个截断点的值是 uint.MaxValue,即 4294967295。显然,这个截断功能只会对"*_STRIP"这类可能无限连续的渲染模式有效。

实际上,glDrawArrays(..)渲染命令可以视为 glDrawElements(..)命令的一个特殊情形,即当 IndexBuffer 的内容为{0, 1, 2, 3, 4, 5, 6, ...}时,使用 glDrawElements(..)命令和使用 glDrawArrays(..)命令的效果是相同的。此时,就可以直接省略这一 IndexBuffer 对象,从而节省资源。进一步讲,在 8.2 章节介绍的拾取算法也是本节介绍的算法的一个特殊情形。

8.3.1 拾取 GL_POINTS 模式下的 Geometry

GL_POINTS 模式下渲染的 PRIMITIVE 是若干 POINT,能够拾取的 Geometry 只有 Point。在 8.1.6 小节中已经介绍了拾取 GL_POINTS 模式下的 Point 的方法,这里不再赘述。要注意的是,使用 DrawElementsCmd 命令时,一个 POINT 可能被多次渲染,所幸这并不影响拾取算法。

8.3.2 拾取 GL_LINES 模式下的 Geometry

GL_LINES 模式下渲染的 PRIMITIVE 是若干 LINE,能够拾取的 Geometry 是 Line 和 Point。如果想拾取 Line,应以 GL_LINES 模式执行 gl_VertexID 渲染方法。由于采用了 flat 关键字,此时无论鼠标位于 LINE 的哪一点上,反推计算得到的 gl_VertexID 都是要拾取的 Line 的最后一个(即第二个)顶点的索引值。注意,此索引值是在 VertexBuffer 中的索引值,即顶点在 VertexBuffer 中的位置。

为了找到 Line 的第一个顶点的索引,需要在 DrawElementsCmd 使用的 IndexBuffer 中查找等于 gl_VertexID 的元素,此元素前面的元素就是 Line 的第一个顶点的索引,如图 8-19 所示。

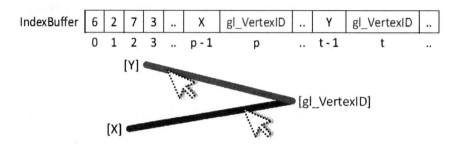

图 8-19　GL_LINES 模式下拾取 Line

图 8-19 中,在 IndexBuffer 里找到 2 个 gl_VertexID 元素,它们之前的元素 X、Y 分别与之构成一个 LINE。这就是说,在模型中存在 2 个 LINE,其第二个顶点在 VertexBuffer 中的索引值都是 gl_VertexID。为了判断鼠标位于哪一个 LINE 上,可以设计一个临时的渲染命令,使之以 GL_LINES 模式渲染{gl_VertexID, X}和{gl_VertexID, Y}组成的 2 个 LINE。注意,这里改变了每个 LINE 的顶点顺序,从而使得 2 个 LINE 的最后一个顶点在 VertexBuffer 中的索引值不同,这样就可以区分出 2 个 LINE。

一般的,在模型中可能存在任意多个 LINE,其第二个顶点在 VertexBuffer 中的索引值都是 gl_VertexID。同样可以用上述方式来区分多个 LINE。

如果想拾取 Point,应以 GL_POINTS 模式执行 gl_VertexID 渲染方法。这就转化为 8.3.1 小节的情形。

8.3.3　拾取 GL_LINE_LOOP 模式下的 Geometry

GL_LINE_LOOP 模式下渲染的 PRIMITIVE 是若干 LINE,能够拾取的 Geometry 是 Line 和 Point。如果想拾取 Line,应以 GL_LINE_LOOP 模式执行 gl_VertexID 渲染方法。由于采用了 flat 关键字,此时无论鼠标位于 LINE 的哪一点上,反推计算得到的 gl_VertexID 都是要拾取的 Line 的最后一个(即第二个)顶点的索引。注意,此索引值是在 VertexBuffer 中的索引值,即顶点在 VertexBuffer 中的位置。

为了找到 Line 的第一个顶点的索引,需要在 DrawElementsCmd 使用的 IndexBuffer 中查找等于 gl_VertexID 的元素,此元素前面的元素就是 Line 的第一个顶点的索引,如图 8-20 所示。

图 8-20 中,在 IndexBuffer 里找到 2 个 gl_VertexID 元素,它们之前的元素 X、Y 分别与之构成一个 LINE。这就是说,在模型中存在 2 个 LINE,其第二个顶点在 VertexBuffer 中的索引值都是 gl_VertexID。为了判断鼠标位于哪一个 LINE 上,可以设计一个临时的渲染命令,使之以 GL_LINES 模式渲染{gl_VertexID, X}和{gl_

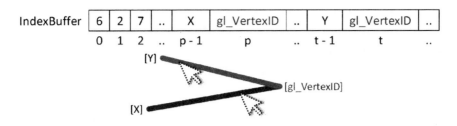

图 8-20 GL_LINE_LOOP 模式下拾取 Line

VertexID,Y}组成的 2 个 LINE。注意,这里改变了每个 LINE 的顶点顺序,从而使得 2 个 LINE 的最后一个顶点在 VertexBuffer 中的索引值不同。这样就可以区分出 2 个 LINE。

一般的,在模型中可能存在任意多个 LINE,其第二个顶点在 VertexBuffer 中的索引值都是 gl_VertexID。同样可以用上述方式来区分多个 LINE。

注意,如果 IndexBuffer 第一个元素的值就是 gl_VertexID,就说明鼠标选择的是由 IndexBuffer 的最后一个元素和第一个元素指定的索引值构成的 LINE,那么就需要特别添加如图 8-21 所示的 LINE 到临时渲染命令。

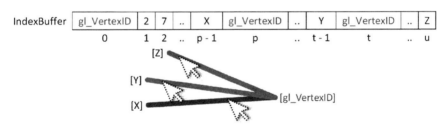

图 8-21 GL_LINE_LOOP 模式下拾取 Line

注意,在 IndexBuffer 中查找到的是{Z, gl_VertexID}这个 LINE;在临时渲染命令中应改为{gl_VertexID, Z},从而保证临时渲染命令中的 LINE 的第二个顶点的索引值各不相同。

如果想拾取 Point,应以 GL_POINTS 模式执行 gl_VertexID 渲染方法。这就转化为 8.3.1 小节的情形。

8.3.4 拾取 GL_LINE_STRIP 模式下的 Geometry

GL_LINE_STRIP 模式下渲染的 PRIMITIVE 是若干 LINE,能够拾取的 Geometry 是 Line 和 Point。此时与拾取 GL_LINES 模式下的 Geometry 情形类似,处理方法完全相同,不再赘述。

8.3.5 拾取 GL_TRIANGLES 模式下的 Geometry

GL_TRIANGLES 模式下渲染的 PRIMITIVE 是若干 TRIANGLE,能够拾取

的 Geometry 是 Triangle、Line 和 Point。

如果想拾取 Triangle，应以 GL_TRIANGLES 模式执行 gl_VertexID 渲染方法。由于采用了 flat 关键字，此时无论鼠标位于 TRIANGLE 的哪一点上，反推计算得到的 gl_VertexID 都是要拾取的 Triangle 的最后一个（即第三个）顶点的索引。

为了找到 Triangle 的前 2 个顶点的索引，需要在 DrawElementsCmd 使用的 IndexBuffer 中查找等于 gl_VertexID 的元素，此元素前面的 2 个元素就是 Triangle 的前 2 个顶点的索引，如图 8-22 所示。

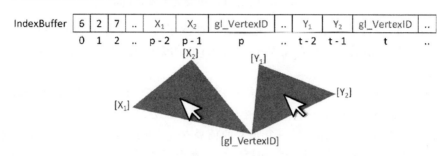

图 8-22　GL_TRIANGLES 模式拾取 Triangle

图 8-22 中，在 IndexBuffer 里找到 2 个 gl_VertexID 元素，它们之前的元素 X_1、X_2 和 Y_1、Y_2 分别与之构成一个 TRIANGLE。这就是说，在模型中存在 2 个 TRIANGLE，其第三个顶点在 VertexBuffer 中的索引值都是 gl_VertexID。为了判断鼠标位于哪一个 TRIANGLE 上，可以设计一个临时的渲染命令，使之以 GL_TRIANGLES 模式渲染{gl_VertexID, X_1, X_2}和{gl_VertexID, Y_1, Y_2}组成的 2 个 TRIANGLE。注意，这里改变了每个 TRIANGLE 的顶点顺序，从而使得 2 个 TRIANGLE 的最后一个顶点在 VertexBuffer 中的索引值不同，这样就可以区分出 2 个 TRIANGLE。

一般的，在模型中可能存在任意多个 TRIANGLE，其第三个顶点在 VertexBuffer 中的索引值都是 gl_VertexID。可以用上述两两对比的方式来逐步找到最终的 TRIANGLE。

如果想拾取 Line，情况就有些复杂了。应启用 glPolygonMode(GL_LINE)，然后以 GL_TRIANGLES 模式执行 gl_VertexID 渲染方法。此时，只有 TRIANGLE 的各个边会被渲染。由于采用了 flat 关键字，当鼠标位于 TRIANGLE 的某一条边上时，反推计算得到的 gl_VertexID 就是 TRIANGLE 的最后一个（即第三个）顶点的索引。

为了找到 Triangle 的前 2 个顶点的索引，需要在 DrawElementsCmd 使用的 IndexBuffer 中查找等于 gl_VertexID 的元素，此元素前面的 2 个元素就是 Triangle 的前 2 个顶点的索引，如图 8-23 所示。

图 8-23 中，在 IndexBuffer 里找到 2 个 gl_VertexID 元素，它们之前的元素 X_1、

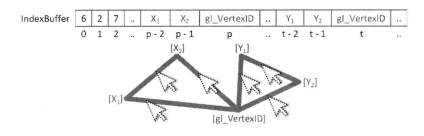

图 8 - 23　GL_TRIANGLES 模式拾取 Line

X_2 和 Y_1、Y_2 分别与之构成一个 TRIANGLE。这就是说,在模型中存在 2 个 TRIANGLE,其第三个顶点在 VertexBuffer 中的索引值都是 gl_VertexID。为了判断鼠标位于哪一个 TRIANGLE 的边上,可以设计一个临时的渲染命令,使之以 GL_TRIANGLES 模式渲染{gl_VertexID, X_1, X_2}和{gl_VertexID, Y_1, Y_2}组成的 2 个 TRIANGLE,并且启用 glPolygonMode(GL_LINE)。注意,这里改变了每个 TRIANGLE 的顶点顺序,从而使得 2 个 TRIANGLE 的最后一个顶点在 VertexBuffer 中的索引值不同。这样就可以区分出要拾取的 Line 在哪个 TRIANGLE。

一般的,在模型中可能存在任意多个 TRIANGLE,其第三个顶点在 VertexBuffer 中的索引值都是 gl_VertexID。可以用上述两两对比的方式来逐步找到最终的 TRIANGLE。

现在,只需在最终的 TRIANGLE 里找到要拾取的 Line 即可,这就转化为 8.3.2 小节中的问题,即设计一个临时渲染命令,以 GL_LINES 模式渲染三个 LINE,从中拾取一个 Line。

如果想拾取 Point,应以 GL_POINTS 模式执行 gl_VertexID 渲染方法。此时,就回到了最简单的拾取情形(即在 GL_POINTS 模式下拾取 Point),处理方法参见 8.3.1 小节。

8.3.6　拾取 GL_TRIANGLE_STRIP 模式下的 Geometry

GL_TRIANGLE_STRIP 模式下渲染的 PRIMITIVE 是若干 TRIANGLE,能够拾取的 Geometry 是 Triangle、Line 和 Point。此时与拾取 GL_TRIANGLES 模式下的 Geometry 情形类似,处理方法完全相同,不再赘述。

8.3.7　拾取 GL_TRIANGLE_FAN 模式下的 Geometry

GL_TRIANGLE_FAN 模式下渲染的 PRIMITIVE 是若干 TRIANGLE,能够拾取的 Geometry 是 Triangle、Line 和 Point。如果想拾取 Triangle,应以 GL_TRIANGLE_FAN 模式执行 gl_VertexID 渲染方法。由于采用了 flat 关键字,此时无论鼠标位于 TRIANGLE 的哪一点上,反推计算得到的 gl_VertexID 都是要拾取的 Triangle 的最后一个(即第三个)顶点的索引。

为了找到 Triangle 的前 2 个顶点的索引,需要在 DrawElementsCmd 使用的 IndexBuffer 中查找等于 gl_VertexID 的元素，此元素前面的 1 个元素就是 Triangle 的前 1 个顶点的索引，而 IndexBuffer 的第一个元素就是 Triangle 的第一个顶点的索引，如图 8-24 所示。

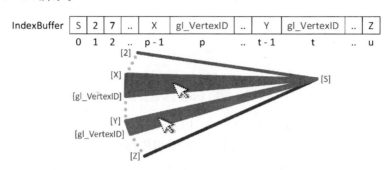

图 8-24　GL_TRIANGLE_FAN 模式下拾取 Triangle

图 8-24 中，在 IndexBuffer 里找到 2 个 gl_VertexID 元素，它们之前的元素 X 和 Y 再加上 S 分别与之构成一个 TRIANGLE。这就是说，在模型中存在 2 个 TRIANGLE，其第三个顶点在 VertexBuffer 中的索引值都是 gl_VertexID。为了判断鼠标在哪一个 TRIANGLE 上，可以设计一个临时的渲染命令，使之以 GL_TRIANGLES 模式渲染{gl_VertexID, S, X}和{gl_VertexID, S, Y}组成的 2 个 TRIANGLE。注意，这里只能以 X 和 Y 作为 TRIANGLE 的最后一个顶点，从而使得 2 个 TRIANGLE 的最后一个顶点在 VertexBuffer 中的索引值不同，这样就可以区分出 2 个 TRIANGLE。

一般的，在模型中可能存在任意多个 TRIANGLE，其第三个顶点在 VertexBuffer 中的索引值都是 gl_VertexID。可以用上述两两对比的方式来逐步找到最终的 TRIANGLE。

如果想拾取 Line，情况就有些复杂。应启用 glPolygonMode(GL_LINE)，然后以 GL_TRIANGLE_FAN 模式执行 gl_VertexID 渲染方法，此时只有 TRIANGLE 的各个边会被渲染。由于采用了 flat 关键字，当鼠标位于 TRIANGLE 的某一条边上时，反推计算得到的 gl_VertexID 就是 TRIANGLE 的最后一个(即第三个)顶点的索引。

为了找到 Triangle 的前 2 个顶点的索引，需要在 DrawElementsCmd 使用的 IndexBuffer 中查找等于 gl_VertexID 的元素，此元素前面的 1 个元素就是 Triangle 的前 1 个顶点的索引，而 IndexBuffer 的第一个元素就是 Triangle 的第一个顶点的索引，如图 8-25 所示。

图 8-25 中，在 IndexBuffer 里找到 2 个 gl_VertexID 元素，它们之前的元素 X 和 Y 再加上 S 分别与之构成一个 TRIANGLE。这就是说，在模型中存在 2 个 TRIANGLE，其第三个顶点在 VertexBuffer 中的索引值都是 gl_VertexID。为了判断鼠标位于哪一个 TRIANGLE 的边上，可以设计一个临时的渲染命令，使之以 GL_

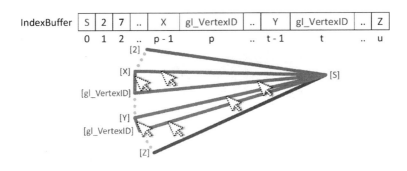

图 8-25　GL_TRIANGLE_FAN 模式下拾取 Line

TRIANGLES 模式渲染{gl_VertexID，S，X}和{gl_VertexID，S，Y}组成的 2 个 TRIANGLE，并且启用 glPolygonMode(GL_LINE)。注意，这里只能以 X 和 Y 作为 TRIANGLE 的最后一个顶点，从而使得 2 个 TRIANGLE 的最后一个顶点在 VertexBuffer 中的索引值不同。这样就可以区分出要拾取的 Line 在哪个 TRIANGLE 上。

一般的，在模型中可能存在任意多个 TRIANGLE，其第三个顶点在 VertexBuffer 中的索引值都是 gl_VertexID。可以用上述两两对比的方式来逐步找到最终的 TRIANGLE。

现在，只需在最终的 TRIANGLE 里找到要拾取的 Line 即可，这就转化为 8.3.2 小节中的问题，即设计一个临时渲染命令，以 GL_LINES 模式渲染三个 Line，从中拾取一个 Line。

如果想拾取 Point，应以 GL_POINTS 模式执行 gl_VertexID 渲染方法。此时就回到了最简单的拾取情形（即在 GL_POINTS 模式下拾取 Point）。处理方法参见 8.1.6 小节。

8.3.8　拾取 GL_QUADS 模式下的 Geometry

GL_QUADS 模式下渲染的 PRIMITIVE 是若干 QUAD，能够拾取的 Geometry 是 Quad、Line 和 Point。如果想拾取 Quad，应以 GL_QUADS 模式执行 gl_VertexID 渲染方法。由于采用了 flat 关键字，此时无论鼠标位于 QUAD 的哪一点上，反推计算得到的 gl_VertexID 都是要拾取的 Quad 的最后一个（即第四个）顶点的索引。

为了找到 Quad 的前 3 个顶点的索引，需要在 DrawElementsCmd 使用的 IndexBuffer 中查找等于 gl_VertexID 的元素，此元素前面的 3 个元素就是 Quad 的前 3 个顶点的索引，如图 8-26 所示。

图 8-26 中，在 IndexBuffer 里找到 2 个 gl_VertexID 元素，它们之前的元素 X_1、X_2、X_3 和 Y_1、Y_2、Y_3 分别与之构成一个 QUAD。这就是说，在模型中存在 2 个 QUAD，其第四个顶点在 VertexBuffer 中的索引值都是 gl_VertexID。为了判断鼠标位于哪一个 QUAD 上，可以设计一个临时的渲染命令，使之以 GL_QUADS 模式

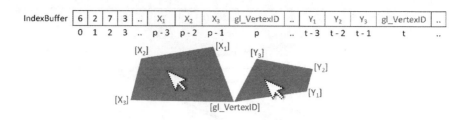

图 8 – 26　GL_QUADS 模式下拾取 Quad

渲染 {gl_VertexID, X_1, X_2, X_3} 和 {gl_VertexID, Y_1, Y_2, Y_3} 组成的 2 个 QUAD。注意，这里改变了每个 QUAD 的顶点顺序，从而使得 2 个 QUAD 的最后一个顶点在 VertexBuffer 中的索引值不同；这样就可以区分出 2 个 QUAD。

一般的，在模型中可能存在任意多个 QUAD，其第四个顶点在 VertexBuffer 中的索引值都是 gl_VertexID。可以用上述两两对比的方式来逐步找到最终的 QUAD。

如果想拾取 Line，情况就有些复杂了。应启用 glPolygonMode(GL_LINE)，然后以 GL_QUADS 模式执行 gl_VertexID 渲染方法，此时只有 QUAD 的各个边会被渲染。由于采用了 flat 关键字，当鼠标位于 QUAD 的某一条边上时，反推计算得到的 gl_VertexID 就是 QUAD 的最后一个（即第四个）顶点的索引。

为了找到 Quad 的前 3 个顶点的索引，需要在 DrawElementsCmd 使用的 IndexBuffer 中查找等于 gl_VertexID 的元素，此元素前面的 3 个元素就是 Quad 的前 3 个顶点的索引，如图 8 – 27 所示。

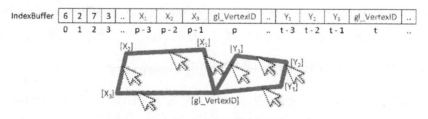

图 8 – 27　GL_QUADS 模式下拾取 Line

图 8 – 27 中，在 IndexBuffer 里找到 2 个 gl_VertexID 元素，它们之前的元素 X_1、X_2、X_3 和 Y_1、Y_2、Y_3 分别与之构成一个 QUAD。这就是说，在模型中存在 2 个 QUAD，其第四个顶点在 VertexBuffer 中的索引值都是 gl_VertexID。为了判断鼠标位于哪一个 QUAD 的边上，可以设计一个临时的渲染命令，使之以 GL_QUADS 模式渲染 {gl_VertexID, X_1, X_2, X_3} 和 {gl_VertexID, Y_1, Y_2, Y_3} 组成的 2 个 QUAD，并且启用 glPolygonMode(GL_LINE)。注意，这里改变了每个 QUAD 的顶点顺序，从而使得 2 个 QUAD 的最后一个顶点在 VertexBuffer 中的索引值不同。这样就可以区分出要拾取的 Line 在哪个 QUAD。

一般的，在模型中可能存在任意多个 QUAD，其第四个顶点在 VertexBuffer 中的索引值都是 gl_VertexID。可以用上述两两对比的方式来逐步找到最终的 QUAD。

现在,只需在最终的 QUAD 里找到要拾取的 Line 即可,这就转化为 8.3.2 小节中的问题,即设计一个临时渲染命令,以 GL_LINES 模式渲染四个 Line,从中拾取一个 Line。

如果想拾取 Point,应以 GL_POINTS 模式执行 gl_VertexID 渲染方法。此时,就回到了最简单的拾取情形(即在 GL_POINTS 模式下拾取 Point)。处理方法参见 8.1.6 小节。

8.3.9　拾取 GL_QUAD_STRIP 模式下的 Geometry

GL_QUAD_STRIP 模式下渲染的 PRIMITIVE 是若干 QUAD,能够拾取的 Geometry 是 Quad、Line 和 Point。此时与拾取 GL_QUADS 模式下的 Geometry 情形类似,处理方法完全相同,不再赘述。

8.3.10　拾取 GL_POLYGON 模式下的 Geometry

GL_POLYGON 模式下渲染的 PRIMITIVE 是若干 POLYGON,能够拾取的 Geometry 是 Polygon、Line 和 Point。如果想拾取 Polygon,应以 GL_POLYGON 模式执行 gl_VertexID 渲染方法。由于采用了 flat 关键字,此时无论鼠标位于 POLYGON 的哪一点上,反推计算得到的 gl_VertexID 都是要拾取的 Polygon 的第一个(这与其他模式不同)顶点的索引。

为了找到 POLYGON 的其他顶点的索引,需要在 DrawElementsCmd 使用的 IndexBuffer 中查找等于 gl_VertexID 的元素,从此元素开始,直到"最后"一个元素,构成了 POLYGON 全部顶点的索引,如图 8-28 所示。

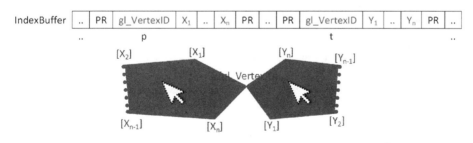

图 8-28　GL_POLYGON 模式下拾取 Polygon

图 8-28 中,在 IndexBuffer 里可能会存在若干个截断点 PR。所谓的"最后"一个元素是指在 gl_VertexID 之后首次出现的 PR 的前一个元素。可以将截断点 PR 的作用视为将一个 IndexBuffer 切割为左右两个独立的 IndexBuffer,这对于 DrawElementsCmd 命令的所有渲染模式下的拾取都适用。

据此,在 IndexBuffer 里找到 2 个 gl_VertexID 元素,它们之后的元素 X_1……X_n 和 Y_1……Y_n 分别与之构成一个 POLYGON。这就是说,在模型中存在 2 个 POLY-

GON，其第一个顶点在 VertexBuffer 中的索引值都是 gl_VertexID。为了判断鼠标在哪一个 POLYGON 上，可以设计一个临时的渲染命令，使之以 GL_POLYGON 模式渲染 $\{X_n, gl\_VertexID, X_1 \cdots\cdots X_{n-1}\}$ 和 $\{Y_n, gl\_VertexID, Y_1 \cdots\cdots Y_{n-1}\}$ 组成的 2 个 POLYGON。注意，这里改变了每个 POLYGON 的顶点顺序，从而使得 2 个 POLYGON 的第一个顶点在 VertexBuffer 中的索引值不同，这样就可以区分出 2 个 POLYGON。

一般的，在模型中可能存在任意多个 POLYGON，其第一个顶点在 VertexBuffer 中的索引值都是 gl_VertexID。可以用上述两两对比的方式来逐步找到最终的 POLYGON。

如果想拾取 Line，情况就有些复杂。应启用 glPolygonMode(GL_LINE)，然后以 GL_POLYGON 模式执行 gl_VertexID 渲染方法。此时，只有 POLYGON 的各个边会被渲染。由于采用了 flat 关键字，当鼠标位于 POLYGON 的某一条边上时，反推计算得到的 gl_VertexID 就是 POLYGON 的第一个顶点的索引（这与其他模式不同）。

为了找到 POLYGON 的其他顶点的索引，需要在 DrawElementsCmd 使用的 IndexBuffer 中查找等于 gl_VertexID 的元素，从此元素开始，直到"最后"一个元素，构成了 POLYGON 全部顶点的索引，如图 8-29 所示。

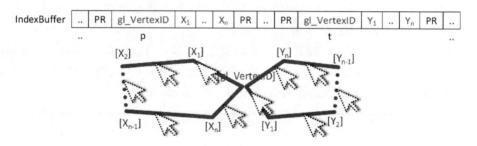

图 8-29　GL_POLYGON 模式下拾取 Line

图 8-29 中，在 IndexBuffer 里可能存在若干个截断点 PR。所谓的"最后"一个元素是指在 gl_VertexID 之后首次出现的 PR 的前一个元素，可以将截断点 PR 的作用视为将一个 IndexBuffer 切割为左右两个独立的 IndexBuffer。这对于 DrawElementsCmd 命令的所有渲染模式下的拾取都适用。

据此，在 IndexBuffer 里找到 2 个 gl_VertexID 元素，它们之后的元素 $X_1 \cdots\cdots X_n$ 和 $Y_1 \cdots\cdots Y_n$ 分别与之构成一个 POLYGON。这就是说，在模型中存在 2 个 POLYGON，其第一个顶点在 VertexBuffer 中的索引值都是 gl_VertexID。为了判断鼠标位于哪一个 POLYGON 上，可以设计一个临时的渲染命令，使之以 GL_POLYGON 模式渲染 $\{X_n, gl\_VertexID, X_1 \cdots\cdots X_{n-1}\}$ 和 $\{Y_n, gl\_VertexID, Y_1 \cdots\cdots Y_{n-1}\}$ 组成的 2 个 POLYGON，并且启用 glPolygonMode(GL_LINE)。注意，这里改变了每个 POLYGON 的顶点顺序，从而使得 2 个 POLYGON 的第一个顶点在 VertexBuffer

中的索引值不同，这样就可以区分出 2 个 POLYGON。

一般的，在模型中可能存在任意多个 POLYGON，其第一个顶点在 VertexBuffer 中的索引值都是 gl_VertexID。可以用上述两两对比的方式来逐步找到最终的 POLYGON。

现在，只需在最终的 POLYGON 里找到要拾取的 Line 即可。这就转化为 8.3.3 小节中的问题，即设计一个临时的渲染命令，使之以 GL_LINE_LOOP 模式渲染这个模型，从中拾取一个 Line。

如果想拾取 Point，应以 GL_POINTS 模式执行 gl_VertexID 渲染方法，此时就回到了最简单的拾取情形（即在 GL_POINTS 模式下拾取 Point）。处理方法参见 8.1.6 小节。

8.3.11　截断的 IndexBuffer

本章使用 PR 代表 IndexBuffer 中的截断点。对于本章介绍的关于 DrawElementsCmd 命令下的拾取，一个 IndexBuffer 中如果存在 PR，就应当视为被 PR 截断为多个 IndexBuffer，然后对截断的各个部分分别执行拾取算法，这是 DrawElementsCmd 命令实现拾取功能的复杂性的原因之一。对于以"_STRIP""_LOOP"或"_FAN"结尾的渲染模式，IndexBuffer 的"第一个"和"最后一个"元素指的是在两个 PR 之间的部分。对于 GL_POLYGON，如果没有 PR，那么整个 IndexBuffer 里所有的数据都在描述同一个（可能很巨大的）POLYGON。对于其他渲染模式，并不需要在 IndexBuffer 中插入 PR。对于所有的渲染模式，可以认为 IndexBuffer 第一个元素之前和最后一个元素之后都存在一个假想的 PR。

如果想要拾取算法能够在任何糟糕的模型数据下都实现拾取操作，那么拾取算法的具体实现逻辑会非常复杂，执行效率也难以提高。另一方面，如果程序中使用了一个糟糕的模型，那就说明开发者应当首先针对糟糕的模型做一些预处理，或者放弃处理它，或者将它打磨得不那么糟糕。因此，CSharpGL 实现的拾取算法只针对干净整齐的模型，不去理会过于糟糕的模型。

那么何谓"干净整齐"的模型？简单来看就是 IndexBuffer 中没有派不上用场的内容。例如，如果一个 GL_TRIANGLES 渲染模式使用的 IndexBuffer 对象中出现了 PR，就是不必要的；如果一个 GL_LINES 渲染模式使用的 IndexBuffer 对象的元素数量为奇数（即缺少一个元素与 IndexBuffer 对象的最后一个元素构成一个 Line），那么最后一个元素就是不必要的；如果 IndexBuffer 中连续两个元素都是 PR，那么第二个 PR 就是不必要的；如果 IndexBuffer 中第一个和最后一个元素是 PR，那么这两个 PR 就是不必要的。

8.4　拖拽顶点

在类似 3D Max 这样的三维模型编辑软件中，用户可以选中要编辑的顶点，然后用鼠标拖拽其到其他位置。本章介绍了单独拾取模型内的某个顶点、线、三角形、四边形和多边形的算法。在拾取的基础上,只需恰当地修改这些顶点的位置属性,就可以实现拖拽顶点的功能。那么,如何"恰当地"修改顶点的位置属性呢?

一般的,记录模型顶点的位置属性的 VertexBuffer,保存的是顶点在 Model Space 中的位置 A。经过 OpenGL Pipeline 渲染过程中的坐标变换,这一位置 A 变为在 Window Space 的位置 B,并显示到画布上。用户通过鼠标点选拾取到的位置,就是在 Window Space 中的位置 B。如果按某个向量来移动鼠标,那么可以视为按相同的向量把位置 B 移动到新的位置 C,且其仍旧位于 Window Space。此时,只需按照坐标变换的逆过程,就可以得到与位置 C 对应的、在 Model Space 中的位置 D。位置 D 与位置 A 的差就是所有被选中的顶点应该移动的向量。据此,即可找到应该在 VertexBuffer 中写入的新的位置。

本小节按照拖拽顶点的流程,介绍如何实现拖拽功能。

首先,当鼠标在画布上移动时,记录下鼠标拾取到的 Geometry 信息。代码如下:

```csharp
partial class FormPicking
{
    private PickedGeometry pickedGeometry; // last picked geometry.

    private void glCanvas1_MouseMove(object sender, MouseEventArgs e)
    {
        if (e.Button == MouseButtons.Left) // if left mouse down.
        {
            // do something.
        }
        else // try picking something.
        {
            int x = e.X;
            int y = this.winGLCanvas1.Height - e.Y - 1;
            this.pickedGeometry = this.pickingAction.Pick(x, y, // picking location.
                PickingGeometryTypes.Triangle,// picking type.
                this.winGLCanvas1.Width,// canvas width.
                this.winGLCanvas1.Height); // canvas height.
        }
    }
}
```

拾取的原理在本章已有详细介绍,不再赘述。注意,此时要在 pickedGeometry 中记录鼠标选中的片段(Fragment)的位置,包括其深度值。如果此时按下鼠标,那么就视为用户想要拖拽当前拾取到的 Geometry,代码如下:

```
partial class FormPicking
{
    private PickedGeometry pickedGeometry; // last picked geometry.
    private DragParam dragParam ; // information for dragging geometry.

    private void glCanvas1_MouseDown(object sender, MouseEventArgs e)
    {
        if (pickedGeometry != null) // if something was picked.
        {
            IGLCanvas canvas = this.winGLCanvas1;
            var viewport = new vec4(0, 0, canvas.Width, canvas.Height);
            var B = new vec3(
                e.X,// x coordinate
                canvas.Height - e.Y - 1,// y coordinate
                pickedGeometry.PickedPosition.z); // z coordinate(depth)
            ICamera camera = this.scene.Camera;
            mat4 projectionMat = camera.GetProjectionMatrix();
            mat4 viewMat = camera.GetViewMatrix();
            mat4 modelMat = (pickedGeometry.FromObject as PickableNode).GetModelMa-
                            trix();
            vec3 A = glm.unProject(
                B , // position at window space.
                viewMat * modelMat, projectionMat, viewport);
                                        // coordinate transformation
            this.dragParam = new DragParam(
                A ,
                projectionMat, viewMat, viewport,
                pickedGeometry.VertexIds);
        }
    }
}
```

按下鼠标后,应记录此时此刻的矩阵、顶点索引等信息,以备后续步骤的对比计算。当鼠标再次移动时,就会发生拖拽现象,代码如下:

```
partial class FormPicking
{
    private PickedGeometry pickedGeometry;
    private DragParam dragParam;

    private void glCanvas1_MouseMove(object sender, MouseEventArgs e)
```

```csharp
            {
                if (e.Button == MouseButtons.Left) // if left mouse down.
                {
                    if (dragParam != null && pickedGeometry != null)
                    {
                        IGLCanvas canvas = this.winGLCanvas1;
                        var node = pickedGeometry.FromObject as PickableNode;
                        var C = new vec3(
                            e.X,// x coordinate
                            canvas.Height - e.Y - 1, // y coordinate
                            pickedGeometry.PickedPosition.z); // z coordinate(depth).
                        vec3 D = glm.unProject(
                            C, // position at window space
                            dragParam.viewMat * node.GetModelMatrix(), // view * model
                            dragParam.projectionMat,// projection
                            dragParam.viewport); // viewport

                        var modelSpacePositionDiff = D - dragParam.A;
                        dragParam.A = D;
                        node.MovePositions(modelSpacePositionDiff, dragParam.pickedVertexIds);
                    }
                }
                else // try picking something
                {
                }
            }
        }
```

随着鼠标的移动，可以找到与之对应的新的位置C。按照 OpenGL Pipeline 的流程，实施一个逆向的流程计算，即将位置C变换为对应的在 Model Space 中的位置D。之后对每个选中的顶点按照D与A的差来移动即可。

修改模型顶点位置的原理十分简单，只需用 glMapBuffer(..)和 glUnmapBuffer(..)来逐个加入D与A的差即可。代码如下：

```csharp
public abstract partial class PickableNode
{
    // Move vertexes' position according to modelSpacePositionDiff(D - A).
```

```
public void MovePositions(vec3 modelSpacePositionDiff, params uint[] positionIn-
        dexes)
{
    VertexBuffer positionBuffer = ..
    IntPtr pointer = positionBuffer.MapBuffer(MapBufferAccess.ReadWrite);
    unsafe
    {
        var array = (vec3 *)pointer.ToPointer();
        foreach (var index in positionIndexes)
        {
            array[index] = array[index] + modelSpacePositionDiff; // add diff
        }
    }
    buffer.UnmapBuffer();
}
```

当放开鼠标时,就视为拖拽动作完毕。此时应清理战场,代码如下:

```
partial class FormPicking
{
    private PickedGeometry pickedGeometry;
    private DragParam dragParam;

    private void glCanvas1_MouseUp(object sender, MouseEventArgs e)
    {
        this.dragParam = null;
    }
}
```

只需清空拖拽参数即可。此时鼠标仍旧位于被拾取的 Geometry 上,因此 pickedGeometry 不应被清空。读者可在本书网络资料上的 Demos\ColorCodedPicking 项目中找到本节的完整代码,并可进行如图 8-30 所示的拖拽操作。

可以看到,拖拽过程中被拾取到的 Geometry 始终跟随着鼠标。模型的某些顶点位置被修改了,使得模型的形状发生改变,这与一般的 3D 建模软件的效果一致。

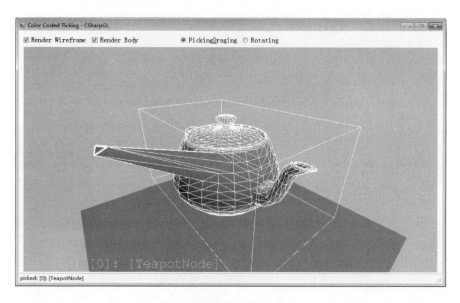

图 8-30　拖拽一个 Triangle

8.5　总　结

在 7.1.1 小节中介绍逻辑操作时已经用到了拾取功能。鼠标移动到哪个 Cube 上，哪个 Cube 就执行逻辑操作，实现反色效果。拾取的目的是能够让用户选择某个模型（或模型上的某个顶点、线段、三角形、四边形或多边形），然后就可以对选中的部分执行各种操作。

本章介绍了 Color-Coded-Picking 算法。这一算法的核心思想是利用 Shader 的内置变量 gl_VertexID 和关键字 flat 实现颜色与索引值的相互转换。由于 OpenGL 支持多种渲染模式和渲染命令，使得这一算法在具体实现时要面对多种可能的情况。因此，需要对整个拾取算法的代码做出完善的设计。

在拾取过程中，如果模型的渲染模式是 GL_POLYGON，那么效率会比较低。应谨慎使用这样的渲染模式。

① 为什么 OpenGL 里能够实现拾取操作？

OpenGL 提供的 GL_SELECT 模式（用于 Legacy OpenGL）、内置变量 gl_VertexID 和 flat 关键字，为拾取操作提供了基础。

② 如何获取 VertexBuffer 中的内容？

用 glMapBuffer(..) 和 glUnmapBuffer()。

③ 如何面对多种渲染模式和渲染命令？

分步处理模型，面向对象设计。

8.6 问 题

带着问题实践是学习 OpenGL 最快的方式。这里给读者提出几个问题,作为抛砖引玉之用。

1. 请设计实验验证,使用 glDrawElements(..)渲染命令,以 GL_POINTS 模式渲染若干个点,那么当 IndexBuffer 中存在重复的索引值时,那个顶点是否会被重复渲染?

2. 请针对所有的渲染模式和渲染命令,分别设计一个典型的模型,然后测试 Color-Coded-Picking 算法。

3. 请尝试用 glReadPixels(..)函数找到鼠标所在位置的深度值。

4. 请尝试用 glReadPixels(..)函数和 glGetTexImage(..)函数将 Texture 对象中的 Image 保存到文件中。

5. 请考虑 Color-Coded-Picking 算法可能的改进思路。

6. 请总结出"干净整齐"的 IndexBuffer 的完整规则。

第9章

文 字

学完本章之后,读者将能:
- 使用 CSharpGL 提供的工具创建自定义的字形图片和配置文件;
- 在三维场景中渲染始终面向 Camera 的文字(Text);
- 实时修改三维场景中的文字内容。

在渲染三维场景时,常常需要在某些模型上添加文字,如图 9-1 所示。

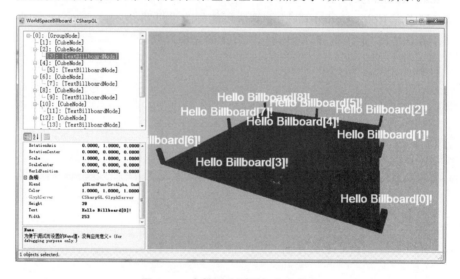

图 9-1　在模型头顶添加文字信息

OpenGL 不支持直接渲染文字。如果需要渲染文字,最基础的思路是把一个个字符做成纹理(Texture)对象,然后把它贴到场景中的适当位置。读者可在本书网络资料中找到 Demos\WorldSpaceBillboard 项目,运行图 9-1 所示的示例程序,观察 9个柱子上的文字随柱子移动的情形。

9.1　固定尺寸且始终面向 Camera

文字应当始终朝向 Camera,且保持固定的尺寸。

9.1.1 恰好铺满全屏的 QUAD

根据 3.2.4 小节,在 NDS(Normalized Device Space)阶段,位于[−1,−1,−1]和[1,1,1]之间的内容才会最终显示到画布上,[−1,−1,−1]和[1,1,1]就是边界线。从 Clip Space 到 NDS,只需将 vec4(x, y, z, w)的各个分量全部除以 w。如果 Clip Space 里的坐标原本就是 vec4(x, y, z, 1),那么连这里的除法操作都可以省略了。换句话说,就是可以在 Clip Space 里直接指定在 NDS 里的坐标,而 Clip Space 里的坐标就是 Vertex Shader 中设定的 gl_Position 变量。因此,只需分别将(±1, ±1, 0, 1)传递给 gl_Position,并且从 Z 轴正半轴向原点方向观察,就可以实现一个恰好覆盖全屏的 QUAD。此 Vertex Shader 代码如下:

```
// vertex shader
#version 150 core

in vec3 inPosition; // position in clip space. (±1, ±1, 0, 1)
in vec2 inUV; // texture coordinate.

out vec2 passUV;

void main() {
    gl_Position = vec4(inPosition, 1.0); // clip space coordinate & NDS coordinate
    passUV = inUV; // pass texture coordinate to fragment shader
}
```

注意,这里直接将 VertexBuffer 中的顶点位置传递给了 gl_Position,而没有经过矩阵变换。这是因为现在已知顶点在 Clip Space 里的位置,且不会发生变化,就可以直接写在 IBufferSource 模型数据里了。此时的模型类型如下:

```
class FullScreenModel : IBufferSource
{
    // positions in clip space and NDS
    private static readonly vec3[] positions = new vec3[]
    {
        new vec3(1, 1, 0), new vec3(-1, 1, 0),
        new vec3(-1, -1, 0), new vec3(1, -1, 0),
    };
    // texture coordinates
    private static readonly vec2[] uvs = new vec2[]
    {
        new vec2(1, 1), new vec2(0, 1),
        new vec2(0, 0), new vec2(1, 0),
```

```csharp
        };

            public const string strPosition = "position";
            private VertexBuffer positionBuffer;
            public const string strUV = "color";
            private VertexBuffer colorBuffer;

            private IDrawCommand drawCmd;

            public IEnumerable<VertexBuffer> GetVertexAttribute(string bufferName)
            {
                if (strPosition == bufferName)
                {
                    if (this.positionBuffer == null)
                    {
                        this.positionBuffer =
                            positions.GenVertexBuffer(VBOConfig.Vec3, BufferUsage.StaticDraw);
                    }

                    yield return this.positionBuffer;
                }
                else if (strUV == bufferName)
                {
                    if (this.colorBuffer == null)
                    {
                        this.colorBuffer = uvs.GenVertexBuffer(VBOConfig.Vec2, BufferUsage.StaticDraw);
                    }

                    yield return this.colorBuffer;
                }
                else
                {
                    throw new ArgumentException();
                }
            }

            public IEnumerable<IDrawCommand> GetDrawCommand()
            {
                if (this.drawCmd == null)
```

```
            {
                this.drawCmd = new DrawArraysCmd(DrawMode.Quads, positions.Length);
            }

            yield return this.drawCmd;
        }
    }
```

读者可在本书网络资料上找到示例项目 c09d00_FullScreenQuad,运行观察始终保持覆盖整个 WinGLCanvas 控件的 QUAD。为了直观地判断出 QUAD 是否恰好覆盖 WinGLCanvas,示例中用图 2-1 所示的贴图铺满此 QUAD,如图 9-2 所示。

图 9-2　始终占据整个画布的 QUAD

9.1.2　固定尺寸的 QUAD

QUAD 的尺寸由用户自己设定,Viewport 的尺寸是已知的,根据 QUAD 的宽高(像素)与 Viewport 的宽高(像素)的比例,就可以求得 NDS 里 QUAD 的坐标。此时在描述 QUAD 的模型中,位置属性可以直接用像素表示,代码如下:

```
class FixedSizeQuadModel : IBufferSource
{
    private vec2[] positions;
    // create a quad with specified width and height in pixels.
    public FixedSizeQuadModel(int width, int height)
    {
        float halfWidth = width / 2.0f;
        float halfHeight = height / 2.0f;
        var positions = new vec2[4];
```

```
            positions[0] = new vec2(halfWidth, halfHeight);
            positions[1] = new vec2(-halfWidth, halfHeight);
            positions[2] = new vec2(-halfWidth, -halfHeight);
            positions[3] = new vec2(halfWidth, -halfHeight);

            this.positions = positions;
        }

        // IBufferSource stuff...
    }
```

相应的 Vertex Shader 代码如下:

```
// vertex shader.
#version 150 core

in vec2 inPosition; // position in pixels.
in vec2 inUV; // texture coordinate.

uniform float screenWidth; // Canvas width.
uniform float screenHeight; //Canvas height.
uniform mat4 mvpMatrix;

out vec2 passUV;

void main() {
vec4 clipSpacePos = mvpMatrix * vec4(0, 0, 0, 1); // move in 3D world.
clipSpacePos = clipSpacePos / clipSpacePos.w; // divided by w.
    clipSpacePos.x += inPosition.x / screenWidth;// move horizontally.
    clipSpacePos.y += inPosition.y / screenHeight;// move vertically.
    gl_Position = clipSpacePos;

    passUV = inUV; // pass texture coordinate to fragment shader.
}
```

Model Space 里的原点(0,0,0,1)就是 QUAD 的中心,因此通过矩阵变换可以得到 QUAD 的中心在 Clip Space 中的位置 clipSpacePos 。在模型 FixedSizeQuad-Model 中,顶点位置描述的是在 Window Space 里的偏移量,这个偏移量就是针对 QUAD 在 Window Space 中的中心位置而言的。由于在 Vertex Shader 中已经将 clipSpacePos.w 的值设定为1,因此它同样是此顶点在 Window Space 中的位置。此 clipSpacePos 加上偏移量占据的画布的比例,就得到了此顶点在 Window Space 中的最终位置。

这里应用的中心思想：无论 QUAD 在三维世界中如何移动、旋转、缩放，其中心位置（即 Model Space 中的原点）相对其父节点都没有改变，因此它是最简单的位置。在将中心位置变换到三维场景中的正确位置后，QUAD 的其他位置就可以通过与中心位置的相对关系来确定，从而保证了 QUAD 既能始终面向 Camera 又能保持固定的尺寸。在应用程序端获取 Viewpoint 尺寸的代码如下：

```
void GetViewportSize(out int width, out int height)
{
    uint target = GL_VIEWPORT;
    var viewport = new int[4];
    glGetIntegerv(target, viewport); // OpenGL function
    width = viewport[2];
    height = viewport[3];
}
```

读者可在本书网络资料上找到示例项目 c09d01_FixedSizeQuad，运行观察始终保持面向 Camera 且尺寸不变的 9 个 QUAD。示例中用被拉高了的 CubeNode 表示要描述的模型，用一个贴图表示 QUAD 并始终保持在 CubeNode 上方，如图 9-3 所示。

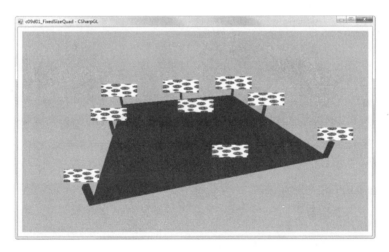

图 9-3 固定尺寸的 QUAD

9.2 字形信息

9.2.1 单独的字形

把需要用到的字符做成一张张小图片，然后紧密地排列到一个大的贴图上，此贴图就可以作为 OpenGL 的 Texture 数据。例如，ASCII 码表中的可见字符可以排列为如图 9-4 所示的一个贴图。

图 9-4　ASCII 码的字形的贴图

图 9-4 中，每个小方框围起来的就是一个字符的字形(Glyph)，第一个是空格。只要将字形贴图的适当位置贴到一个 QUAD 上，就实现了渲染相应的字符的功能。使用 C♯ 提供的下述函数测量一个字符或字符串的尺寸：

```
// C# code
public sealed class Graphics
{
    // measure size of text in specified font
    public SizeF MeasureString(string text, Font font);
}
```

但是这个函数返回的结果不是 text 本身的宽度，而是包含了在左右两侧自动添加的空白的宽度。为了获取单个字符 x 的宽度，可以用字符串 "| x |" 的宽度减去字符串 "| |" 的宽度。这样就得到了 x 本身的宽度。使用 C♯ 提供的下述函数可以将一个字符或字符串画到 Bitmap 上：

```
// C# code
public sealed class Graphics
{
    // draw string with specified font , brush and location(x, y) to bitmap
    public SizeFDrawString(string s, Font font, Brush brush, float x, float y);
}
```

但是此函数会像 MeasureString(..) 函数一样给指定的字符串左右添加空白，这将使得最终的贴图不够紧密，会浪费大量的空间。因此应用下述函数将字符的字形以一个小 Bitmap 的方式画到最终的贴图上：

```
// C# code
public sealed class Graphics
{
    // draw image in specified rect to specified rect of destination image
    publicvoid DrawImage(Image image, RectangleF destRect, RectangleF srcRect,
        GraphicsUnit srcUnit);
}
```

通过简单的实验可知，MeasureString(..)和 DrawString(..)函数自动给字符串添加的空白部分，左侧和右侧的空白宽度是相等的，因此可以先用 DrawString(..)分别得到字符串"｜x｜"和"｜｜"的 Image，然后再用 DrawImage(..)将 "｜x｜" 的 Image 中 x 的部分画到最终的贴图上。

9.2.2 字形纹理信息

一个贴图上有很多个字形，每个字形都需要记录自己在贴图中的 UV 坐标。例如，包含了 ASCII 码和中文字符的字形服务器里的字形如图 9-5 所示。注意，由于中文字符数量太多，图中只展示了其中 6 页的内容。左上方的是第一个贴图，右下方的是最后一个贴图。这些贴图大小都相同，可以将其存放到一个二维数组纹理（2D Array Texture）对象中，每个贴图的 Page 序号就是它在纹理中的索引值。

图 9-5 字形服务器中的部分贴图

将需要的字符依次安排到合适的位置，这一任务由下述代码实现：

```csharp
// C# code
class StringChunk
{
    public string Text { get; private set; }
    public Font TheFont { get; private set; }

    // Put(..) method will fill these properties
    public SizeF Size { get; set; }
    public PointF LeftTop { get; set; }
    public int PageIndex { get; set; }

    // a string to be put on image
    public StringChunk(string text, Font font)
        : base(text, font)
    {
        this.Text = text;
        this.TheFont = font;
    }

    // put this string to image
    public void Put(PagesContext context)
    {
        if (context == null) { throw new ArgumentNullException("context"); }
        Graphics graphics = context.UnitGraphics;
        if (graphics == null) { throw new Exception("Context already disposed!"); }
        // no more pages allowed.
        if (context.CurrentIndex >= context.PageList.Count) { return; }

        Page page = context.PageList[context.CurrentIndex];
        PointF leftTop = context.CurrentLeftTop;
        SizeF bigSize = graphics.MeasureString("|" + this.Text + "|", this.The-
                        Font);
        SizeF emptySize = graphics.MeasureString("||", this.TheFont);
        var width = (bigSize.Width - emptySize.Width); // width of this string.
        var height = bigSize.Height; // height of this string.
        if (page.Width < leftTop.X + width) // time to next line.
        {
            // time to next page
            if (page.Height < leftTop.Y + context.MaxLineHeight + height)
```

```
                {
                    context.CurrentIndex ++ ;
                    if (context.PageList.Count <= context.CurrentIndex
                                                              // this page is full
                        && context.PageList.Count < context.MaxPageCount)
                                                              // new page available
                    {
                        var newPage = new Page(context.PageWidth, context.PageHeight);
                        context.PageList.Add(newPage);
                    }

                    // use next page whether it exists or not
                    {
                        this.LeftTop = new PointF(0, 0);
                        this.Size = new SizeF(width, height);
                        this.PageIndex = context.CurrentIndex;

                        context.CurrentLeftTop = new PointF(width, 0);
                        context.MaxLineHeight = height;
                    }
                }
                else // next line, not next page
                {
                    leftTop = new PointF(0, leftTop.Y + context.MaxLineHeight);
                    this.LeftTop = leftTop;
                    this.Size = new SizeF(width, height);
                    this.PageIndex = context.CurrentIndex;

                    context.CurrentLeftTop = new PointF(leftTop.X + width, leftTop.Y);
                    context.MaxLineHeight = height;
                }
            }
            else // put this string in current line
            {
                this.LeftTop = leftTop;
                this.Size = new SizeF(width, height);
                this.PageIndex = context.CurrentIndex;

                context.CurrentLeftTop = new PointF(leftTop.X + width, leftTop.Y);
                if (leftTop.X == 0 // this is the first string in the line
                    || context.MaxLineHeight < height)
```

```
                {
                    context.MaxLineHeight = height;
                }
            }
        }
    }
}
```

将各个字符或字符串封装为 StringChunk 对象,依次执行 Put(..)方法,即可获得对应字形在贴图上的位置。据此即可绘制贴图,并得到每个字形在贴图上的 UV 坐标。

9.3 三维世界的文字

三维世界中的文字,例如三维游戏中人物角色的头顶上显示的名字,是始终朝向 Camera 且保持大小不变的。根据本章介绍的内容,只需将字形贴图的适当位置贴到一个个 QUAD 的表面,即可渲染出这样的文字。

本节以创建 ASCII 码字形库为例,简要介绍创建字形服务器的过程。一个字形服务器,包含字形贴图、UV 坐标信息和一个字典(用于查找可用的字形)。创建 ASCII 码字形服务器的基本流程代码如下:

```
// C# code
GlyphServer CreateASCIIServer()
{
    var builder = new StringBuilder();
    // ASCII characters
    for (char c = ' '; c <= '~'; c ++) // from space(' ') to ('~')
    {
        builder.Append(c);
    }
    string charSet = builder.ToString();
    float emSize = 64.0f;
    var font = new System.Drawing.Font("Arial", emSize, GraphicsUnit.Pixel);
    int maxTextureWidth = 1024, maxTextureHeight = 1024, maxPageCount = 1000;
    GlyphServer server = GlyphServer.Create(font, charSet,
        maxTextureWidth, maxTextureHeight, maxPageCount);
    font.Dispose();

    return server;
}
```

```
GlyphServer Create(Font font, IEnumerable<char> charset, int maxTextureWidth, int
        maxTextureHeight, int maxPageCount)
{
    if (charset == null || charset.Count() == 0) { return new GlyphServer(); }
    // encapsulate each character into a chunk.
    List<StringChunk> chunkList = GetChunkList(font, charset);
    // context records states of pages
    var context = new PagesContext(maxTextureWidth, maxTextureHeight, maxPageCount);
    foreach (var item in chunkList)
    {
        item.Put(context); // put a character into pages
    }
    // generates bitmaps that are just big enough for all chunks
    Bitmap[] bitmaps = GenerateBitmaps(chunkList, context.PageList.Count);
    // print all chunks into bitmaps according to coordinates got from Put(..)
    PrintChunks(chunkList, context, bitmaps);
    // record every character and its UV information in the dictionary
    // generate a GL_TEXTURE_2D_ARRAY texture object with bitmaps
    Texture texture = GenerateTexture(bitmaps);
    Dictionary<string, GlyphInfo> dictionary = GetDictionary(chunkList, bitmaps[0].Width, bitmaps[0].Height);

    var server = new GlyphServer(); // glyph server object keeps everything above
    server.dictionary = dictionary;
    server.GlyphTexture = texture;
    server.TextureWidth = bitmaps[0].Width;
    server.TextureHeight = bitmaps[0].Height;

    foreach (var item in bitmaps) // dispose bitmap objects
    {
        item.Dispose();
    }

    return server;
}
```

创建一个字形服务器,需要输入的信息包括:字符 charSet、文字格式 Font 和贴图最大尺寸。

首先，将单个的字符封装为StringChunk，它包含字符本身、字符使用的文本格式Font以及尚未安排好的位置信息。接着根据贴图的尺寸信息，逐一地将StringChunk的位置安排好，每个StringChunk的位置信息由自己内部的字段保存。然后，根据所有StringChunk的位置信息，可以找到恰好能将其包含在内的Bitmap尺寸，据此生成若干Bitmap对象。注意，OpenGL支持的Texture对象的尺寸是有限的，所以如果使用的字形数量很多（例如想要渲染中文字符），就应该考虑将StringChunk分别放到多个Bitmap中。

根据所有StringChunk的位置信息，用Graphics.DrawImage(..)等方法将其字形画到上一步准备好的Bitmap[]中的适当位置。然后，将已经准备好了的Bitmap[]作为Image数据，创建一个二维数组纹理对象（详情请参考2.2.2小节）。此纹理对象能够保存多个Bitmap的内容，最适于这里的情况。接着创建一个字典对象dictionary，用于记录此字形服务器都拥有哪些字符。当用户需要使用某个字符时，可以从字典对象中迅速查到相关信息，最后将字典对象、纹理对象及其他信息全部保存到GlyphServer对象，返回即可。注意，此过程中创建的Image[]已经不再使用，此处应当显式地调用Dispose()方法将其释放掉。

如果想使用除ASCII码外的字符，只需在创建GlyphServer时将需要使用的字符传入构造函数。例如，需要使用中文时，构造GlyphServer的代码如下：

```
// C# code.
GlyphServer CreateChineseServer()
{
    var builder = new StringBuilder();
    // ASCII characters…
    // Chinese characters.
    for (char c = (char)0x4E00; c <= 0x9FA5; c++) from ('一') to ('龥')
    {
        builder.Append(c);
    }
    string charSet = builder.ToString();
    //…


    return server;
}
```

那么示例项目Demos\WorldSpaceBillboard可显示如图9-6所示的内容。

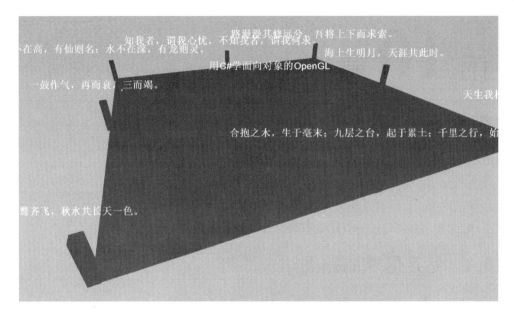

图 9-6 显示中文文字和符号

9.4 总　结

本章介绍了一种在三维世界中渲染文字的方法。

① 渲染文字的思路是什么？

把一个字形贴到一个 QUAD 上。

② 如何修改文本？

用 glMapBuffer(..) 和 glUnmapBuffer()。

9.5 问　题

带着问题实践是学习 OpenGL 最快的方式。这里给读者提出几个问题，作为抛砖引玉之用。

1. 请设计实验验证 DrawString(..) 方法两侧的空白宽度是否相同？

2. 请设计实验验证 DrawString(..) 方法绘制的字符串 "｜x｜" 的高度与字符串 "｜｜" 的高度是否相同？

第 10 章

简单的用户界面

学完本章之后，读者将能：
- 在 OpenGL 中设计实现简单的用户界面；
- 为控件（GLControl）设置布局属性。

10.1 指定区域的贴图

如果能在三维场景的最前方提供一些用户界面（UI,User Interface），可以极大地丰富用户的交互途径。在图 7-8 中，窗口右上方被一个贴图占据，用于显示当前使用的 Shadow Mapping 的贴图内容。整个程序渲染的场景在三维世界中，此贴图以固定的尺寸和位置展示在二维的画布前方，这是本节介绍的固定尺寸的贴图的一个示例。在窗口指定的位置以指定的尺寸将一些内容渲染到最前方（从而呈现为二维 UI 的样式），是渲染控件（Control）的基础。

在 9.1.1 和 9.1.2 小节中已经介绍了如何渲染铺满全屏的 QUAD 和三维场景中固定尺寸的 QUAD。为了实现固定尺寸的 QUAD，需要在 Vertex Shader 或 IBufferSource 中计算复杂的位置信息，因此并不适于渲染二维 UI。在 3.2.5 小节中介绍了 glViewport(..)函数的功能，在 7.1.1 小节中介绍了裁剪测试功能。在此基础上，可以避免复杂的位置计算，并且将渲染范围限制在指定的区域内，代码如下：

```
//region: vec4(x, y, width, height), background: vec4(r, g, b, a)
void RenderGUI(vec4 region, vec4 background, ..)
{
    glEnable(GL_SCISSOR_TEST); // enable scissor test
    glScissor(region.x, region.y, region.z, region.w); // set up scissor region
    glViewport(region.x, region.y, region.z, region.w); // set up viewport region

    glClearColor(background.x, background.y, background.z, background.w);
                                                      // set clear color
    glClear(GL_COLOR_BUFFER_BIT); // clear region with clear color

    // render UI
}
```

glViewport(..)可以将大多数渲染结果填充到指定的区域，但是对于类似 gl-Clear(..)这样的功能是无效的。因此需要用裁剪测试将所有渲染操作都限制到指定的区域。如果在指定的区域内不使用 glClear(GL_COLOR_BUFFER_BIT); ，那么在此区域内的背景色就会保持画布原本的背景色，这可以用来实现渲染不规则形状的控件。

10.2 控件的布局机制

CSharpGL 支持的控件布局机制，遵循下述设计原则：
➢ 各个控件被组织为树结构；
➢ 一个控件的布局属性的改变，只会影响其子控件的布局属性。

各个控件组织成的树结构的根节点（Root Control）负责记录和更新画布（IGLCanvas）的尺寸。由根节点和 2 个子节点组织成的一个示例如图 10-1 所示。

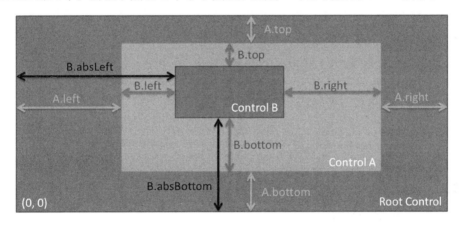

图 10-1 由 3 个 Control 组织成的一个树结构

如图 10-1 中的 Control B 所示，每个控件都记录自己到父节点的上边距（top）、下边距（bottom）、左边距（left）和右边距（right）。此外，每个控件都记录自己的宽度（Width）和高度（Height）。显然，父节点的宽度等于子节点的"left+Width+right"；父节点的高度等于子节点的"bottom+Height+top"。为了避免每次渲染时都重复计算，每个控件还记录了自身到根节点的左边距（absLeft）和下边距（absBottom）。

控件的另一重要属性 Anchor，描述了控件的哪一边锁定了与父控件的距离。可以用一个枚举类型实现 Anchor，代码如下：

```csharp
// C# code.
[Flags]
public enum GUIAnchorStyles
{
    None = 0,
    Top = 1,
    Bottom = 2,
    Left = 4,
    Right = 8,
}
public partial class GLControl
{
        private static readonly GUIAnchorStyles leftAnchor = GUIAnchorStyles.Left;
        private static readonly GUIAnchorStyles rightAnchor = GUIAnchorStyles.Right;
        private static readonly GUIAnchorStyles bottomAnchor = GUIAnchorStyles.Bottom;
        private static readonly GUIAnchorStyles topAnchor = GUIAnchorStyles.Top;
        private static readonly GUIAnchorStyles leftRightAnchor =
            GUIAnchorStyles.Left | GUIAnchorStyles.Right;
        private static readonly GUIAnchorStyles bottomTopAnchor =
            GUIAnchorStyles.Bottom | GUIAnchorStyles.Top;
}
```

可见 Anchor 属性完全照搬了 C# 库的设计。为书写方便，在所有控件的基类 GLControl 中添加了常用的 Anchor 组合常量。

10.2.1 宽度联动

父节点的宽度改变时，仅会影响子节点的 left、Width、right 或 absLeft，代码如下：

```csharp
public partial class GLControl
{
    private int width;
    // Width of this control
    public int Width
    {
        get { return this.width; }
        set
        {
            if (this.width != value)
            {
```

```
                this.width = value; // update width field
                GLControl parent = this.parent;
                if (parent != null)
                {
                    this.right = parent.width - this.left - value;
                }
                GLControl.LayoutAfterWidthChanged(this, this.Children);
            }
        }
    }
    // update layout property of this control's children
    static void LayoutAfterWidthChanged(GLControl parent, GLControlChildren glControlChildren)
    {
        foreach (var control in glControlChildren)
        {
            GUIAnchorStyles anchor = control.Anchor;
            if ((anchor & leftRightAnchor) == leftRightAnchor)
                                                            // lock left and right
            { control.Width = parent.width - control.left - control.right; }
                                                            // update Width
            else if ((anchor & leftAnchor) == leftAnchor)    // lock left only
            { control.right = parent.width - control.left - control.width; }
                                                            // update right
            else if ((anchor & rightAnchor) == rightAnchor)  // lock right only
            {
                control.left = parent.width - control.right - control.width;
                                                            // update left
                control.absLeft = parent.absLeft + control.left; // update absLeft
                foreach (var item in control.Children)  // update absLeft of children
                {
                    UpdateAbsLeft(control, item);
                }
            }
            else // if ((anchor & noneAnchor) == noneAnchor) // lock nothing
            {   // put control in middle of parent
                int diff = parent.width - control.left - control.width - control.right;
                int halfDiff = diff / 2;
                intotherHalfDiff = diff - halfDiff;
```

```
                control.left += halfDiff;
                control.absLeft = parent.absLeft + control.left;
                control.right += otherHalfDiff;
                foreach (var item in control.Children) // update absLeft in children
                {
                    UpdateAbsLeft(control, item);
                }
            }
        }
    }
}
// update absLeft field recursively
static void UpdateAbsLeft(GLControl parent, GLControl control)
{
    control.absLeft = parent.absLeft + control.left;

    foreach (var item in control.Children)
    {
        UpdateAbsLeft(control, item);
    }
}
```

当 GLControl 的 Width 属性改变时,除了修改控件自身的 width 和 right 字段外,还要修改每个子控件在水平方向上的布局属性值。子控件与父控件的绑定关系不同,子控件的布局属性值的修改方式也就相应的不同了。

10.2.2 高度联动

父节点的高度改变时,仅会影响子节点的 bottom、Height、top 或 absBottom,代码如下:

```
public partial class GLControl
{
    private int height;
    // Height of this control
    public int Height
    {
        get { return this.height; }
        set
        {
            if (this.height != value)
            {
                this.height = value;
```

```
                        GLControl parent = this.parent;
                        if (parent != null)
                        {
                            this.top = parent.height - this.bottom - value;
                        }
                        GLControl.LayoutAfterHeightChanged(this, this.Children);
                    }
                }
            }
            // update layout property of this control's children
            static void LayoutAfterHeightChanged(GLControl parent, GLControlChildren glControlChildren)
            {
                foreach (var control in glControlChildren)
                {
                    GUIAnchorStyles anchor = control.Anchor;
                    if ((anchor & bottomTopAnchor) == bottomTopAnchor)
                    { control.Height = parent.height - control.bottom - control.top; }
                    else if ((anchor & bottomAnchor) == bottomAnchor)
                    { control.top = parent.height - control.bottom - control.height; }
                    else if ((anchor & topAnchor) == topAnchor)
                    {
                        control.bottom = parent.height - control.top - control.height;
                        control.absBottom = parent.absBottom + control.bottom;
                    }
                    else// if ((anchor & noneAnchor) == noneAnchor)
                    {
                        int diff = parent.height - control.top - control.height - control.bottom;
                        int halfDiff = diff / 2;
                        int OtherHalfDiff = diff - halfDiff;
                        control.bottom += halfDiff;
                        control.absBottom = parent.absBottom + control.bottom;
                        control.top += OtherHalfDiff;
                    }
                }
            }
            // update absBottom field recursively
            static void UpdateAbsBottom(GLControl parent, GLControl control)
            {
                control.absBottom = parent.absBottom + control.bottom;
```

```
        foreach (var item in control.Children)
        {
            UpdateAbsBottom(control, item);
        }
    }
}
```

当 GLControl 的 Height 属性改变时,除了修改控件自身的 height 和 top 字段外,还要修改每个子控件在竖直方向上的布局属性值。子控件与父控件的绑定关系不同,子控件的布局属性值的修改方式也就相应的不同了。

10.2.3 左边距联动

父节点的左边距(X)改变时,仅会影响子节点的 absLeft,代码如下:

```
public partial class GLControl
{
    private int left;
    // left distance to parent
    public int X
    {
        get { return this.left; }
        set
        {
            if (this.left != value)
            {
                this.left = value;
                GLControl parent = this.parent;
                if (parent != null)
                {
                    this.right = parent.width - value - this.width;
                    GLControl.UpdateAbsLeft(parent, this);
                }
            }
        }
    }
}
```

改变 X 的含义是改变控件的位置,不改变控件的尺寸。

10.2.4 下边距联动

父节点的下边距(Y)改变时,仅会影响子节点的 absBottom,代码如下:

```
public partial class GLControl
{
    private int bottom;
    // Left Down location
    public int Y
    {
        get { return this.bottom; }
        set
        {
            if (this.bottom != value)
            {
                this.bottom = value;
                GLControl parent = this.parent;
                if (parent != null)
                {
                    this.top = parent.height - value - this.height;
                    GLControl.UpdateAbsBottom(this, parent);
                }
            }
        }
    }
}
```

改变 Y 的含义是改变控件的位置,不改变控件的尺寸。

10.2.5　父节点联动

控件的父控件(Parent)被替换时,要通知原来的父控件消除对自己的记录,要让新的父控件记录自己,要根据自己对父控件的绑定关系,重新计算自己的布局属性值。由于新的父控件可能与原来的父控件完全不同,所以要重新计算所有子控件的绝对位置:absLeft 和 absBottom。代码如下:

```
public partial class GLControl
{
    internal GLControl parent; // GLControlChildren needs to use this field directly
    // Parent control.
    public GLControl Parent
    {
        get { return this.parent; }
        set
        {
            GLControl old = this.parent;
            if (old != value)
```

```
                {
                    this.parent = value;// this records the parent

                    if (old != null)
                    {
                        old.Children.Remove(this); // bye, old parent
                    }

                    if (value != null)
                    {
                        value.Children.Add(this); // the parent records this

                        GLControl.LayoutAfterAddChild(this, value);
                        GLControl.UpdateAbsLocation(this, value);
                    }
                }
            }

        static void LayoutAfterAddChild(GLControl control, GLControl parent)
        {
            GUIAnchorStyles anchor = control.Anchor;
            if ((anchor & leftRightAnchor) == leftRightAnchor)
            { control.Width = parent.width - control.left - control.right; }
            else if ((anchor & leftAnchor) == leftAnchor)
            { control.right = parent.width - control.left - control.width; }
            else if ((anchor & rightAnchor) == rightAnchor)
            { control.left = parent.width - control.width - control.right; }
            else// if ((anchor & noneAnchor) == nonAnchor)
            {
                int diff = parent.width - control.left - control.width
                        - control.right;
                int halfDiff = diff / 2;
                int otherHalfDiff = diff - halfDiff;
                control.left += halfDiff;
                control.right += otherHalfDiff;
            }

            if ((anchor & bottomTopAnchor) == bottomTopAnchor)
            { control.Height = parent.height - control.bottom - control.top; }
            else if ((anchor & bottomAnchor) == bottomAnchor)
```

```
            { control.top = parent.height - control.bottom - control.height; }
            else if ((anchor & topAnchor) == topAnchor)
            { control.bottom = parent.height - control.height - control.top; }
            else// if ((anchor & noneAnchor) == nonAnchor)
            {
                int diff = parent.height - control.bottom - control.height - control.top;
                int halfDiff = diff / 2;
                int otherHalfDiff = diff - halfDiff;
                control.bottom += halfDiff;
                control.top += otherHalfDiff;
            }
        }
    }
```

注意,在让新的父控件记录自己时(value.Children.Add(this);),并没有考虑到 Children 内部会再次请求父控件记录自己的情形(即引起反复循环调用),这将在下一小节解决。

10.2.6 子节点联动

控件用一个列表类型 GLControlChildren 来记录自己的所有子节点(Children)。此类型在添加新的控件时,会请求新控件也记录自己代表的控件为父控件,这将与上一小节的 Parent 属性的行为构成无限的循环调用。为了避免这一情况,可以将 GL-Control 的 parent 字段和 GLControlChildren 的 children 字段的访问权限都设置为 internal,即整个 CSharpGL 库内部都可以访问。这样,在设置 Parent 属性时,Parent 可以用 value.Children.children.Add(this); 代替 value.Children.Add(this); 在使用 Add(..)方法时,可以用 item.parent = this.parent; 代替 item.Parent = this.parent; ,从而避免反复相互调用:

```
// children in GLControl
public class GLControlChildren : IList<GLControl>
{
    // GLControl needs to use this field directly.
    internal readonly List<GLControl> children = new List<GLControl>();

    // parent of this list's items
    private readonly GLControl parent;

    // add item to tail of this list
    public void Add(GLControl item)
    {
```

```
            if (item == null) { return; }

            item.parent = this.parent; // item's parent records this
            children.Add(item);

            GLControl.LayoutAfterAddChild(item, this.parent);
            GLControl.UpdateAbsLocation(item, this.parent);
        }
}
```

对于 GLControlChildren 中每个添加成员的方法，都需要注意避免反复相互调用的情况，同时也要更新布局属性；但在删除成员时，只需直接删除。

10.3 控件的事件机制

控件树的根节点除了维护窗口的尺寸外，还可以负责控件的事件机制。根控件响应画布的鼠标和键盘事件，其响应的方式：根据具体情况，找到某个 GLControl，让它来响应此事件。

在 WinForm 中有对鼠标和键盘事件的定义，在其他平台中也有各自对鼠标和键盘事件的定义。OpenGL 本身是一个标准，每个平台都可以有自己的 OpenGL 实现。因此，CSharpGL 也应该只定义一个关于鼠标和键盘事件的标准，而不是直接使用 WinForm 对鼠标和键盘事件的定义。

```
// generic event handler of mouse or keyboard event
public delegate void GLEventHandler<TEventArgs>(object sender, TEventArgs e);
```

这个定义是完全仿照 WinForm 的定义设计的，连同相关的参数类型等的定义都是完全仿照 WinForm 的定义设计的。看似画蛇添足，但是能够让 CSharpGL 不依赖具体的平台。每个平台都只负责实现自己的根节点控件。WinForm 下的根节点控件代码如下：

```
// Root Control in CSharpGL's scene's GUI system
public class WinCtrlRoot : CtrlRoot
{
    // binding canvas
    public WinGLCanvas BindingCanvas { get; private set; }

    // mouse and keyboard event handler
    private readonly System.Windows.Forms.MouseEventHandler mouseMove;
    private readonly System.Windows.Forms.MouseEventHandler mouseDown;
    private readonly System.Windows.Forms.MouseEventHandler mouseUp;
```

```csharp
        private readonly System.Windows.Forms.KeyEventHandler keyDown;
        private readonly System.Windows.Forms.KeyEventHandler keyUp;
        private readonly System.EventHandler resize;

        // current control that we are working on
        private GLControl currentControl;

        public WinCtrlRoot(int width, int height)
            : base(width, height)
        {
            this.mouseMove = new MouseEventHandler(winCanvas_MouseMove);
            this.mouseDown = new MouseEventHandler(winCanvas_MouseDown);
            this.mouseUp = newMouseEventHandler(winCanvas_MouseUp);
            this.keyDown = newKeyEventHandler(winCanvas_KeyDown);
            this.keyUp = new KeyEventHandler(winCanvas_KeyUp);
            this.resize = new EventHandler(winCanvas_Resize);
        }
        // bind this control to specified canvas
        public override void Bind(IGLCanvas canvas)
        {
            var winCanvas = canvas as WinGLCanvas;
            if (winCanvas == null) { throw new ArgumentException(); }

            winCanvas.MouseMove += mouseMove;
            winCanvas.MouseDown += mouseDown;
            winCanvas.MouseUp += mouseUp;
            winCanvas.KeyDown += keyDown;
            winCanvas.KeyUp += keyUp;
            winCanvas.Resize += resize;

            this.BindingCanvas = winCanvas;
        }
        // update control's size
        void winCanvas_Resize(object sender, EventArgs e)
        {
            var control = sender as System.Windows.Forms.Control;
            this.Width = control.Width;
            this.Height = control.Height;
        }

        void winCanvas_KeyUp(object sender, System.Windows.Forms.KeyEventArgs e)
```

```csharp
        {
            if (this.currentControl != null)
            {
                GLKeyEventArgs args = e.Translate();
                this.currentControl.InvokeEvent(EventType.KeyUp, args);
            }
        }

        void winCanvas_KeyDown(object sender, System.Windows.Forms.KeyEventArgs e)
        {
            if (this.currentControl != null)
            {
                GLKeyEventArgs args = e.Translate();
                this.currentControl.InvokeEvent(EventType.KeyDown, args);
            }
        }

        void winCanvas_MouseUp(object sender, System.Windows.Forms.MouseEventArgs e)
        {
            if (this.currentControl != null)
            {
                GLMouseEventArgs args = e.Translate();
                this.currentControl.InvokeEvent(EventType.MouseUp, args);
            }
        }

        void winCanvas_MouseDown(object sender, System.Windows.Forms.MouseEventArgs e)
        {
            int x = e.X, y = this.BindingCanvas.Height - e.Y - 1;
            GLControl control = GetControlAt(x, y, this);
            this.currentControl = control;
            if (control != null)
            {
                GLMouseEventArgs args = e.Translate();
                control.InvokeEvent(EventType.MouseDown, args);
            }
        }

        // Get the control at specified position
        private GLControl GetControlAt(int x, int y, GLControl control)
```

```
    {
        GLControl result = null;
        if (control.ContainsPoint(x, y))
        {
            if (control.AcceptPicking)
            {
                result = control;
            }

            foreach (var item in control.Children)
            {
                GLControl child = GetControlAt(x, y, item);
                if (child != null)
                {
                    result = child;
                    break;
                }
            }
        }

        return result;
    }

    void winCanvas_MouseMove(object sender, System.Windows.Forms.MouseEventArgs e)
    {
        if (this.currentControl != null)
        {
            GLMouseEventArgs args = e.Translate();
            this.currentControl.InvokeEvent(EventType.MouseMove, args);
        }
    }
    // unbind this control to current canvas
    public override void Unbind()
    {
        WinGLCanvas winCanvas = this.BindingCanvas;
        if (winCanvas != null)
        {
            winCanvas.MouseMove -= mouseMove;
            winCanvas.MouseDown -= mouseDown;
            winCanvas.MouseUp -= mouseUp;
            winCanvas.KeyDown -= keyDown;
```

```
            winCanvas.KeyUp -= keyUp;

            this.BindingCanvas = null;
        }
    }
}
```

当使用 Bind(..)方法将根控件 WinCtrlRoot 绑定到指定的 WinGLCanvas 后，WinCtrlRoot 就会收到 WinGLCanvas 的鼠标和键盘事件消息，然后让当前的 GLControl 去处理这个消息。

类似 WinForm 的控件，每个 GLControl 都有自己的鼠标和键盘事件的成员。当根节点通知自己要处理某个消息时，Control 就会根据参数来调用相应的事件成员，代码如下：

```
public abstract partial class GLControl
{
    public event GLEventHandler<GLKeyEventArgs> KeyDown;
    public event GLEventHandler<GLKeyEventArgs> KeyUp;
    public event GLEventHandler<GLMouseEventArgs> MouseDown;
    public event GLEventHandler<GLMouseEventArgs> MouseUp;
    public event GLEventHandler<GLMouseEventArgs> MouseMove;

    public virtual void InvokeEvent(EventType eventType, GLEventArgs args)
    {
        switch (eventType)
        {
        case EventType.KeyDown:
            var keyDown = this.KeyDown;
            if (keyDown != null)
            {
                var e = args as GLKeyEventArgs;
                keyDown(this, e);
            }
            break;
        case EventType.KeyUp:
            var keyUp = this.KeyUp;
            if (keyUp != null)
            {
                var e = args as GLKeyEventArgs;
                keyUp(this, e);
            }
            break;
```

```
                case EventType.MouseDown:
                    var mouseDown = this.MouseDown;
                    if (mouseDown != null)
                    {
                        var e = args as GLMouseEventArgs;
                        mouseDown(this, e);
                    }
                    break;
                case EventType.MouseUp:
                    var mouseUp = this.MouseUp;
                    if (mouseUp != null)
                    {
                        var e = args as GLMouseEventArgs;
                        mouseUp(this, e);
                    }
                    break;
                case EventType.MouseMove:
                    var mouseMove = this.MouseMove;
                    if (mouseMove != null)
                    {
                        var e = args as GLMouseEventArgs;
                        mouseMove(this, e);
                    }
                    break;
                default:
                    throw new NotDealWithNewEnumItemException(typeof(EventType));
            }
        }
```

综上所述，WinCtrlRoot 起到一个中转站的作用，将 WinForm 的消息传递给某个 GLControl。只要以 WinCtrlRoot 为根节点，就可以让整个控件系统具有自动调整布局和处理 WinForm 消息的功能。如果有相应的根节点的实现，换做其他平台也是适用的。

10.4 CtrlImage

本节介绍一个 CSharpGL 提供的控件 CtrlImage，它可以将一个 Image 展示在指定的区域内，作用类似于一个 PictureBox，CtrlImage 直接继承自 GLControl，它需要的模型数据很简单，只要一个边长为 2、中心在原点的正方形即可，代码如下：

```csharp
class CtrlImageModel : IBufferSource
{
    private static readonly vec2[] positions = new vec2[]
    { new vec2(1, 1), new vec2(-1, 1), new vec2(-1, -1), new vec2(1, -1), };
    private static readonly vec2[] uvs = new vec2[]
    { new vec2(1, 1), new vec2(0, 1), new vec2(0, 0), new vec2(1, 0), };

    public const string strPosition = "position";
    private VertexBuffer positionBuffer;

    public const string strUV = "uv";
    private VertexBuffer uvBuffer;

    private IDrawCommand drawCmd;

    #region IBufferSource members

    public IEnumerable<VertexBuffer> GetVertexAttribute(string bufferName)
    {
        if (bufferName == strPosition)
        {
            if (this.positionBuffer == null)
            {
                this.positionBuffer = positions.GenVertexBuffer(
                    VBOConfig.Vec2, BufferUsage.StaticDraw);
            }

            yield return this.positionBuffer;
        }
        else if (bufferName == strUV)
        {
            if (this.uvBuffer == null)
            {
                this.uvBuffer = uvs.GenVertexBuffer(
                    VBOConfig.Vec2, BufferUsage.StaticDraw);
            }

            yield return this.uvBuffer;
        }
        else
        {
```

```
            throw new ArgumentException();
        }
    }

    public IEnumerable<IDrawCommand> GetDrawCommand()
    {
        if (this.drawCmd == null)
        {
            this.drawCmd = new DrawArraysCmd(DrawMode.Quads, positions.Length);
        }

        yield return this.drawCmd;
    }

    #endregion
}
```

可见 CtrlImage 只需画一个 QUAD，并且把贴图贴在上面即可。其 Shader 代码如下：

```
// vertex shader.
#version 330 core

in vec3 inPosition;
in vec2 inUV;

out vec2 passUV;

void main() {
    gl_Position = vec4(inPosition, 1.0);
    passUV = inUV;
}

// -------------------------------------------------
// fragment shader.
#version 330 core

in vec2 passUV;
uniform sampler2D tex;

out vec4 outColor;

void main() {
    vec4 color = texture(tex, passUV);
    outColor = color;
}
```

CtrlImage 本身的其他部分与 Scene 里的节点类似,只有在渲染时有所不同,代码如下:

```
public partial class CtrlImage : GLControl
{
    public override void RenderGUIBeforeChildren(GUIRenderEventArgs arg)
    {
        base.RenderGUIBeforeChildren(arg); // scissor test, viewport, background

        ModernRenderUnit unit = this.RenderUnit;
        RenderMethod method = unit.Methods[0];
        method.Render();
    }
}
```

渲染 GLControl 时不再需要找 Camera 设置参数,因为 GLControl 本身不属于常规的三维场景,它的位置是与 Camera 无关的。

10.5　CtrlLabel

本节介绍一个 CSharpGL 提供的控件 CtrlLabel,它可以显示一行文字,作用类似于 WinForm 中的 Label。CtrlLabel 直接继承自 GLControl,它能够显示一个 GlyphServer 包含的所有字符。类似于三维场景中的 TextBillboardNode,它通过将每个字符的位置和 UV 坐标依次安排到一行中,实现渲染一行文字的功能。不同的是,它显示的文字呈现在三维场景之前,其位置由布局属性决定。修改 CtrlLabel 的文本,代码如下:

```
public partial class CtrlLabel
{
    private string text = string.Empty;
    private GlyphsModel labelModel;
    private VertexBuffer positionBuffer;
    private VertexBuffer stringBuffer;
    private DrawArraysCmd drawCmd;

    // text changed event
    public event EventHandler TextChanged;

    protected void DoTextChanged()
    {
        var textChanged = this.TextChanged;
```

```csharp
            if (textChanged != null)
            {
                TextChanged(this, EventArgs.Empty);
            }
        }
        // text property
        public unsafe string Text
        {
            get { return text; }
            set
            {
                string v = value == null ? string.Empty : value;
                if (v != text)
                {
                    text = v;
                    ArrangeCharaters(v, GlyphServer.DefaultServer);
                    DoTextChanged();
                }
            }
        }

        unsafe private void ArrangeCharaters(string text, GlyphServer server)
        {
            float totalWidth, totalHeight;
            PositionPass(text, server, out totalWidth, out totalHeight);
            UVPass(text, server);

            this.drawCmd.VertexCount = text.Length * 4;
                                                    // each alphabet needs 4 vertexes

            this.Width = (int)(totalWidth * this.Height / totalHeight);
                                                    // auto size means auto width
        }
        // start from (0, 0) to the right
        unsafe private void UVPass(string text, GlyphServer server)
        {
            VertexBuffer buffer = this.stringBuffer;
            var textureIndexArray = (QuadSTRStruct *)buffer.MapBuffer(MapBufferAccess.WriteOnly);
            int index = 0;
            foreach (var c in text)
```

```csharp
            {
                if (index >= this.labelModel.Capacity) { break; }

                GlyphInfo glyphInfo;
                if (server.GetGlyphInfo(c, out glyphInfo)) // find glyph info in the dict
                {
                    textureIndexArray[index] = glyphInfo.quad;
                }

                index ++ ;
            }

            buffer.UnmapBuffer();
        }
        // start from (0, 0) to the right
        unsafe private void PositionPass(string text, GlyphServer server,
            out float totalWidth, out float totalHeight)
        {
            int textureWidth = server.TextureWidth;
            int textureHeight = server.TextureHeight;
            VertexBuffer buffer = this.positionBuffer;
            var positionArray = (QuadPositionStruct *) buffer.MapBuffer(MapBufferAccess.ReadWrite);
            const float height = 2.0f; // let's say height is 2.0f
            totalWidth = 0;
            totalHeight = height;
            int index = 0;
            foreach (var c in text)
            {
                if (index >= this.labelModel.Capacity) { break; }

                GlyphInfo glyphInfo;
                float wByH = 0;
                if (server.GetGlyphInfo(c, out glyphInfo))
                {
                    float w = (glyphInfo.quad.rightBottom.x - glyphInfo.quad.leftBottom.x) * textureWidth;
                    float h = (glyphInfo.quad.rightBottom.y - glyphInfo.quad.rightTop.y) * textureHeight;
                    wByH = height * w / h;
                }
```

```csharp
        else
        {
            // put an empty glyph(square) here.
            wByH = height * 1.0f / 1.0f;
        }

        var leftTop = new vec2(totalWidth, height / 2);
        var leftBottom = new vec2(totalWidth, - height / 2);
        var rightBottom = new vec2(totalWidth + wByH, - height / 2);
        var rightTop = new vec2(totalWidth + wByH, height / 2);
        positionArray[index ++ ] = new QuadPositionStruct(leftTop, leftBottom, rightBottom, rightTop);
        totalWidth += wByH;
    }

    // move to center
    for (int i = 0; i < text.Length; i ++ )
    {
        if (i >= this.labelModel.Capacity) { break; }

        QuadPositionStruct quad = positionArray[i];
        var newPos = new QuadPositionStruct(
            // y is already in [-1, 1], so just shrink x to [-1, 1].
            new vec2(quad.leftTop.x / totalWidth * 2.0f - 1f, quad.leftTop.y),
            new vec2(quad.leftBottom.x / totalWidth * 2.0f - 1f, quad.leftBottom.y),
            new vec2(quad.rightBottom.x / totalWidth * 2.0f - 1f, quad.rightBottom.y),
            new vec2(quad.rightTop.x / totalWidth * 2.0f - 1f, quad.rightTop.y));

        positionArray[i] = newPos;
    }

    buffer.UnmapBuffer();
}
```

注意，glMapBuffer(..)和 glUnmapBuffer()函数必须成对使用。如果在使用 glMapBuffer(..)读写一个 Buffer 时，又使用 glMapBuffer(..)试图读写另一个 Buffer 的话，会引发错误，因此这里只能将对位置的修改和对 UV 的修改分开完成。最终的位置坐标被规范到了[-1，-1]和[1，1]之间，这就保证了 Vertex Shader 中

得到的是期望的Clip Space中的坐标,代码如下:

```glsl
// vertex shader
#version 330 core

in vec2 inPosition; // position in clip space
in vec3 inSTR;

out vec3 passSTR;

void main() {
    gl_Position = vec4(inPosition, 0.0, 1.0);
    passSTR = inSTR;
}

// ---------------------------------------------------
// fragment shader
#version 330 core

in vec3 passSTR;

uniform sampler2DArray glyphTexture;
uniform vec3 textColor;

out vec4 outColor;

void main() {
    float a = texture(glyphTexture, vec3(passSTR.xy, floor(passSTR.z))).a;
    outColor = vec4(textColor, a);
}
```

注意,由于GlyphServer里保存的字形可能由多个大小相同的Image保存,因此这里使用了sampler2DArray类型的取样器,UV坐标也进化为由三个分量描述的STR坐标,其中最后一个坐标描述的是二维数组纹理中Image的索引值。

10.6 CtrlButton

本节介绍CSharpGL提供的一个控件CtrlButton,它可以显示一个按钮,作用类似于WinForm中的Button。Button自身可以用2个QUAD实现从上到下的渐变色,以凸显其立体的样式。Button上的文字可以直接用一个CtrlLabel实现,方法是添加一个CtrlLabel作为其子节点,代码如下:

```csharp
public partial class CtrlButton : GLControl
{
    protected override void DoInitialize()
    {
        this.RenderUnit.Initialize();

        var label = new CtrlLabel(100, GUIAnchorStyles.None);
        //label.RenderBackground = true // for debug purpose
        label.Initialize();
        label.AcceptPicking = false;
                            // this label should never be picked by event system.
        this.label = label;

        this.Children.Add(label);

        label.Text = "Button";
        label.TextChanged += label_TextChanged;
    }

    void label_TextChanged(object sender, EventArgs e)
    {
        CtrlLabel label = this.label;
        // move label to center
        {
            int diffX = this.Width - label.Width;
            int diffY = this.Height - label.Height;
            label.Location = new GUIPoint(diffX / 2, diffY / 2);
        }
    }
}
```

读者可以在本书网络资料上的 c10d00_UserInterface 项目中找到 CtrlImage、CtrlLabel 和 CtrlButton 的示例，如图 10-2 所示。

图 10-2 中，左上角的三个控件分别为 CtrlLabel、CtrlButton 和 CtrlImage。

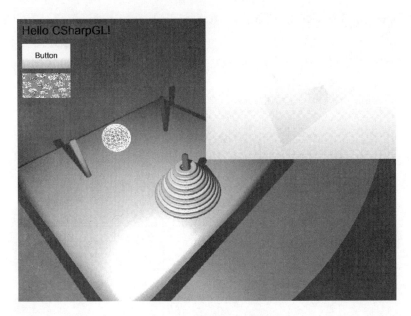

图 10-2 在场景上放置控件

10.7 总 结

本章介绍了一种在三维世界中渲染用户界面的方法。介绍了 CSharpGL 使用的控件布局机制和事件响应机制。

① 如何设计控件布局机制？

控件组织为树结构，节点布局属性的改变只影响子节点。

② 如何设计事件响应机制？

统一的事件类型，从属于各个平台的根节点控件。

10.8 问 题

带着问题实践是学习 OpenGL 最快的方式。这里给读者提出几个问题，作为抛砖引玉之用。

1. 请设计实验验证 GLControl 的布局机制对各种 Anchor 下的反应是否都与 WinForm 相同？

2. 请在 Shader 中实现 CtrlLabel 中文字的倾斜、加粗、下划线等功能。

第11章 轨迹球

学完本章之后,读者将能:
- 了解用轨迹球(Arcball)操纵模型或 Camera 的原理;
- 实现和使用轨迹球。

11.1 轨迹球

在观察模型时,用户常常需要旋转模型,从而能从各个角度观察它。轨迹球就是一种将用户的鼠标操作转换为对模型的旋转命令的技术,它能够让用户以一种直观的方式旋转模型。所谓的轨迹球,是在画布上假想的一个半球,它恰好覆盖整个画布,如图 11-1 所示。

图 11-1 (左)在画布里渲染一个简单的场景 (右)画布上方的假想半球即为轨迹球

轨迹球的基本原理如下:

当鼠标在画布的某点 A 按下时,在假想的轨迹球上有对应的一点 B,使得 AB 连线垂直于画布所在的平面。然后,当鼠标移动到画布的某点 C 时,在假想的轨迹球上同样有对应的一点 D,使得 CD 连线垂直于画布所在的平面。分别用 B 和 D 连接轨迹球的球心 O,那么,从向量 OB 到向量 OD 的旋转操作,就应该是施加到模型上的旋转操作,这一过程如图 11-2 所示。

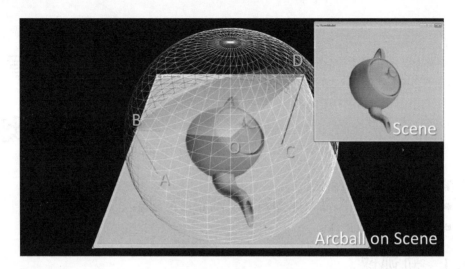

图 11-2 轨迹球工作原理

读者可打开本书网络资料的 c11d00_Arcball 项目,详细观察在鼠标按下并拖动时轨迹球上发生的情况。

11.2 使用

用面向对象的观点,可以设计一个 ArcballManipulater 类型,用于将用户的鼠标操作转换为相应的旋转矩阵。注意,当用户完成一次鼠标的"按下—移动—弹起"操作后,ArcballManipulater 要记录此次生成的旋转矩阵。这样,当用户再次完成"按下—移动—弹起"操作后,ArcballManipulater 就可以通过矩阵相乘的方式得到将两次操作合并的旋转矩阵。当用户继续实施"按下—移动—弹起"操作时,也同样地可以将多次操作合并,这就是矩阵乘法的方便之处。

ArcballManipulater 对象的创建和用法示例如下:

```
private void FormMain_Load(object sender, EventArgs e)
{
    //...

    // create an arcball manipulater object who uses left mouse click
    var manipulater = new ArcBallManipulater(GLMouseButtons.Left);
    // bind it to camera and canvas so that it starts to work
    manipulater.Bind(camera, this.winGLCanvas1);
    // update model's rotation when manipulater decides to rotate
    manipulater.Rotated += manipulater_Rotated;
}
```

```
void manipulater_Rotated(object sender, ArcBallManipulater.Rotation e)
{
    // update node's rotation angle and axis according to manipulater
    SceneNodeBase node = this.scene.RootNode;
    if (node != null)
    {
        node.RotationAngle = e.angleInDegree;
        node.RotationAxis = e.axis;
    }
}
```

注意,要想使用 ArcballManipulater 对象,必须让它绑定到需要的摄像机和画布上,ArcballManipulater 对象内部会自动监听画布上的鼠标事件,从而实现记录旋转矩阵的功能。如果一个节点渲染的模型需要服从 ArcballManipulater 对象的旋转操作,它就应该注册一个 Rotated 事件,在事件中更新自己的旋转轴和旋转角度。这样,ArcballManipulater 对象的启用与否可以通过绑定和取消绑定来实现,ArcballManipulater 对象能影响到的节点也由节点自己来决定。ArcballManipulater 对象与节点之间实现了一个简单的观察者模式。

11.3 设 计

创建 ArcballManipulater 对象时要初始化一些内容,代码如下:

```
public partial class ArcBallManipulater : Manipulater, IMouseHandler
{
    private bool isBinded = false;

    public GLMouseButtons BindingMouseButtons { get; set; }

    private readonly GLEventHandler<GLMouseEventArgs> mouseDownEvent;
    private readonly GLEventHandler<GLMouseEventArgs> mouseMoveEvent;
    private readonly GLEventHandler<GLMouseEventArgs> mouseUpEvent;

    public ArcBallManipulater(GLMouseButtons bindingMouseButtons = GLMouseButtons.Left)
    {
        this.BindingMouseButtons = bindingMouseButtons;
        // creates event objects
        this.mouseDownEvent = ((IMouseHandler)this).canvas_MouseDown;
        this.mouseMoveEvent = ((IMouseHandler)this).canvas_MouseMove;
        this.mouseUpEvent = ((IMouseHandler)this).canvas_MouseUp;
    }
}
```

绑定到指定的画布的含义就是让 ArcballManipulater 对象的鼠标的"按下—移动—弹起"函数注册到画布的鼠标事件，代码如下：

```csharp
public partial class ArcBallManipulater
{
    // Bind this manipulater to specified camera and canvas
    public override void Bind(ICamera camera, IGLCanvas canvas)
    {
        this.camera = camera;
        this.canvas = canvas;

        canvas.MouseDown += this.mouseDownEvent;
        canvas.MouseMove += this.mouseMoveEvent;
        canvas.MouseUp += this.mouseUpEvent;

        this.isBinded = true;
    }
}
```

当鼠标按下时，除了初始化一些变量外，最主要的工作是计算和记录鼠标所在位置 A 对应到轨迹球上的位置 B，代码如下：

```csharp
public partial class ArcBallManipulater
{
    void IMouseHandler.canvas_MouseDown(object sender, GLMouseEventArgs e)
    {
        if ((e.Button & this.BindingMouseButtons) != GLMouseButtons.None)
        {
            // records canvas' size
            IGLCanvas canvas = this.canvas;
            this.SetBounds(this.canvas.Width, this.canvas.Height);

            // get position at arcball surface relative to mouse position on window
            vec3 startPos = GetArcBallPosition(e.X, e.Y);
            this._startPosition = startPos;

            mouseDownFlag = true;

            {   // invoke mouse down event
                var mouseDown = this.MouseDown;
                if (mouseDown != null)
                {
```

```
                mouseDown(this, e);
            }
        }
    }
}
// get position at arcball surface relative to mouse position on window
private vec3 GetArcBallPosition(int x, int y)
{
    float rx = (x - _width / 2.0f) / _length;
    float ry = (_height / 2.0f - y) / _length;
    float zz = _radiusRadius - rx * rx - ry * ry;
    float rz = (zz > 0 ? (float)Math.Sqrt(zz) : 0.0f);
    // one way to do it
    var result = new vec3(
        rx * _vectorRight.x + ry * _vectorUp.x + rz * _vectorBack.x,
        rx * _vectorRight.y + ry * _vectorUp.y + rz * _vectorBack.y,
        rx * _vectorRight.z + ry * _vectorUp.z + rz * _vectorBack.z
        );
    // another way to do it
    //var position = newvec3(rx, ry, rz); // position in model space
    //var matrix = new mat3(_vectorRight, _vectorUp, _vectorBack);
    //result = matrix * position;

    return result;
}
// records window's size
private void SetBounds(int width, int height)
{
    this._width = width; this._height = height;
    _length = width > height ? width : height; // longer length to cover all window
    var rx = ((float)(width) / 2) / _length;
    var ry = ((float)(height) / 2) / _length;
    _radiusRadius = (float)(rx * rx + ry * ry); // cache (radius * radius)
}
```

函数 GetArcBallPosition(int x，int y)用于获取鼠标所在位置对应到轨迹球上的位置。ArcballManipulater 对象为了能够处理在任意摄像机下的轨迹球，需要将轨迹球上的位置变换到 Camera Space。从式(3-9)可以看到，将轨迹球上的位置变换到 Camera Space 里的位置，所需要的变换矩阵可以直接由 Camera 的参数得出，即式(11-1)。

$$\begin{bmatrix} right.x & right.y & right.z \\ up.x & up.y & up.z \\ back.x & back.y & back.z \end{bmatrix} \qquad (11-1)$$

轨迹球的球心是定义在 Model Space 的原点的,轨迹球在 World Space 也无需变动位置,因此可以省略 Model Matrix 的变换过程。因此 vec3(rx, ry, rz)可以直接与 View Matrix 相乘,从而得到在 Camera Space 下的坐标。

当鼠标移动时,要计算鼠标所在位置 C 对应到轨迹球上的位置 D,然后根据从 B 到 D 的旋转来计算旋转矩阵,代码如下:

```csharp
public partial class ArcBallManipulater
{
    void IMouseHandler.canvas_MouseMove(object sender, GLMouseEventArgs e)
    {
        if (mouseDownFlag && ((e.Button & this.lastBindingMouseButtons) != GLMouse-
            Buttons.None))
        {
            this._endPosition = GetArcBallPosition(e.X, e.Y);
            var cosRadian = this._startPosition.dot(this._endPosition) / (this._st-
                        artPosition.length() * this._endPosition.length());
            float angle = (float)(Math.Acos(cosRadian) / Math.PI * 180);
            this._normalVector = this._startPosition.cross(this._endPosition).nor-
                        malize();
            if (!
                ((_normalVector.x == 0 && _normalVector.y == 0 && _normalVector.z
                            == 0)
                || float.IsNaN(_normalVector.x)
                || float.IsNaN(_normalVector.y)
                || float.IsNaN(_normalVector.z)))
            {
                this._startPosition = this._endPosition;
                // get new total rotation by multiplying matrix
                mat4 newRotation = glm.rotate(angle,this._normalVector);
                this.totalRotation = newRotation * this.totalRotation;

                GLEventHandler<Rotation> rotated = this.Rotated;
                if (rotated != null) // invoke rotate event
                {
                    Quaternion quaternion = this.totalRotation.ToQuaternion();
                    float angleInDegree;
                    vec3 axis;
                    quaternion.Parse(out angleInDegree, out axis);
```

```
                    rotated(this, new Rotation(axis, angleInDegree,
                        e.Button, e.Clicks, e.X, e.Y, e.Delta));
                }
            }
        }
    }
}
```

注意,此时鼠标在轨迹球上已经发生了移动,因此也应该相应地旋转模型,即调用 Rotate 事件,这就实现了实时通知所有注册了的节点更新旋转状态。

11.4 四元数

三维世界里的旋转(rotate),可以用一个 3×3 的矩阵描述,也可以用"旋转角度+旋转轴"描述。数学家欧拉证明了这两种形式可以相互转化,且多次旋转可以归结为一次旋转,这实际上就是轨迹球算法旋转模型的理论基础。

四元数就是一个四维向量(w, x, y, z),其中 w 描述旋转的角度(但不是直接的 angleDegree 值),(x, y, z)描述旋转轴。从 angleDegree 和 axis 得到一个四元数是比较简单的,代码如下:

```
public struct Quaternion
{
    private float w;
    private float x;
    private float y;
    private float z;

    // Quaternion from a rotation axis and angle in degree
    public Quaternion(float angleDegree, vec3 axis)
    {
        vec3 normalized = axis.normalize();
        double radian = angleDegree * Math.PI / 180.0;
        double halfRadian = radian / 2.0;
        this.w = (float)Math.Cos(halfRadian);
        float sin = (float)Math.Sin(halfRadian);
        this.x = sin * normalized.x;
        this.y = sin * normalized.y;
        this.z = sin * normalized.z;
    }
}
```

从上面的定义可以很容易推算出四元数里蕴含的 angleDegree 和 axis，代码如下：

```
public void Parse(out float angleDegree, out vec3 axis)
{
    angleDegree = (float)(Math.Acos(w) * 2 * 180.0 / Math.PI);
    axis = (new vec3(x, y, z)).normalize();
}
```

11.5 总　结

本章介绍了轨迹球的概念、用法和设计。
① 轨迹球是什么？
位于窗口上的假想的半球。
② 如何累加历次旋转操作？
用矩阵乘法。

11.6 问　题

带着问题实践是学习 OpenGL 最快的方式。这里给读者提出问题，作为抛砖引玉之用。

请在本书网络资料上的 c11d00_Arcball 项目中，在 FormModel 窗口上旋转模型，并在 FormArcball 窗口中观察对应的轨迹球上的旋转情况。试分析此项目都使用了哪些有趣的技术？

第12章

体渲染

学完本章之后,读者将能:
- 了解体渲染(Volume Rendering)的特点和用途;
- 用 Raycast 方式实现体渲染;
- 用传统的混合功能实现体渲染。

12.1 什么是体渲染

通常用 OpenGL 渲染模型,实际上渲染的是模型的表面。例如,渲染一个立方体时,实际上只渲染了立方体的 6 个面,这个立方体是空心的。而体渲染关注的是模型内部的每个位置上的属性值,即渲染实心的模型。例如,在医学成像中,通过专业机器可以扫描得到患者整个头部的核磁共振数据,体渲染就可以将这些数据渲染出来,便于医生诊断病情。此时要渲染的不是患者头部的外表面,而是整个内部数据。体渲染的示例如图 12-1 所示。

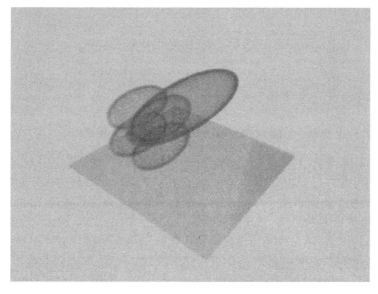

图 12-1 用体渲染技术渲染三维数据

本章介绍几种实现体渲染的方法。

12.2 静态切片

利用 OpenGL 的混合功能可以实现顺序相关的半透明渲染。利用 3D Texture 对象可以方便地存储体数据。如果将一个实心的立方体切为很多片，并且按照从远到近的顺序渲染，就可以得到一个简单的体渲染效果。请读者在本书网络资料上找到 c12d00_StaticSlices 项目，运行后可以看到运用这一思路渲染体数据的效果。当观察者（摄像机）的位置合适时，能够观察到由多个切片混合构成的体渲染效果，如图 12-2 所示。

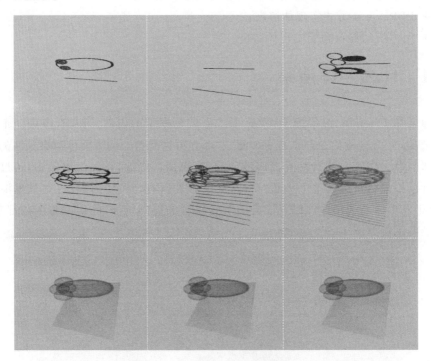

图 12-2　静态切片方法实现体渲染

图 12-2 中，按从上到下、从左到右的顺序依次展示了切片数量为 1、2、4、8、16、32、64、128 和 256 时的渲染结果。OpenGL 用一个 3D Texture 对象来记录完整的体数据，每个切片上的顶点都有对应处的纹理坐标。在此角度观察时，远处的切片先被渲染，近处的切片后被渲染，在混合模式为 glBendFunc(GL_SRC_ALPHA，GL_ONE_MINUS_SRC_ALPHA) 时，就可以观察到期望的结果。但若调整摄像机的位置（或者旋转模型），使得远处的切片后被渲染，渲染效果就会逐渐消失。

静态切片方法实际上是在渲染若干个并列的 QUAD，如图 12-3 所示。

体渲染

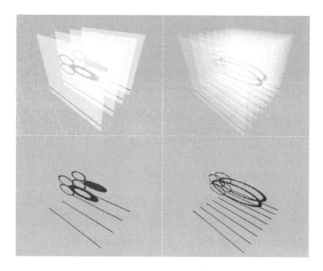

图 12-3 （上）纯色的 QUAD （下）根据纹理着色的 QUAD

图 12-3 中，上方分别展示了 4 个和 8 个 QUAD 并列且混合渲染的效果，下方对应地展示了 4 个和 8 个根据三维纹理着色的 QUAD 并列且混合渲染的效果，这说明静态切片方法不需要特别为体数据创建模型，而是可以直接使用一个简单且统一的立方体切片模型。只需将其他扫描数据上传到三维纹理对象，就可以用同一模型展示其他的数据。静态切片使用的 Vertex Shader 和 Fragment Shader 代码十分简单，代码如下：

```
// vertex shader.
#version 150

in vec3 inPosition; // position
in vec3 inTexCoord; // 3D texture coordinate
out vec3 passTexCoord;

uniform mat4 mvpMat; // projection * view * model

void main() {
    gl_Position = mvpMat * vec4(inPosition, 1.0);
    passTexCoord = inTexCoord;
}

// --------------------------------------------------
//fragment shader
```

```glsl
#version 150
in vec3 passTexCoord;

uniform sampler1D tex1D; // 1D texture for color
uniform sampler3D tex3D; // 3D texture for value

out vec4 outColor;

void main() {
    float x = texture(tex3D, passTexCoord).x;
    outColor = texture(tex1D, x);
}
```

静态切片方法需要的切片模型也十分简单,代码如下:

```csharp
class StaticSlicesModel : IBufferSource
{
    public StaticSlicesModel(int sliceCount)
    {
        if (sliceCount < 1) { throw new ArgumentOutOfRangeException(); }

        var positions = new vec3[sliceCount * 4];
        for (int i = 0; i < sliceCount; i++)
        {
            positions[i * 4 + 0] = new vec3((float)i / (float)(sliceCount - 1), 1, 1);
            positions[i * 4 + 1] = new vec3((float)i / (float)(sliceCount - 1), 0, 1);
            positions[i * 4 + 2] = new vec3((float)i / (float)(sliceCount - 1), 0, 0);
            positions[i * 4 + 3] = new vec3((float)i / (float)(sliceCount - 1), 1, 0);
        }
        positions.Move2Center();
        this.positions = positions;

        var texCoords = new vec3[sliceCount * 4];
        for (int i = 0; i < sliceCount; i++)
        {
            texCoords[i * 4 + 0] = new vec3((float)i / (float)(sliceCount - 1), 1, 1);
            texCoords[i * 4 + 1] = new vec3((float)i / (float)(sliceCount - 1), 0, 1);
            texCoords[i * 4 + 2] = new vec3((float)i / (float)(sliceCount - 1), 0, 0);
            texCoords[i * 4 + 3] = new vec3((float)i / (float)(sliceCount - 1), 1, 0);
        }
        this.texCoords = texCoords;
    }
}
```

这一模型初始时占据着[0,0,0]到[1,1,1]的空间,在Move2Center()函数处理过后,会占据[-0.5,-0.5,-0.5]到[0.5,0.5,0.5]的空间,即中心在原点边长为1的立方体范围。注意,模型初始占据的[0,0,0]到[1,1,1]空间时的顶点位置属性值与顶点纹理坐标属性值是完全相同的,这是立方体的一个特殊性质。如果用到Shader中,可以减少所需的VertexBuffer数量和长度。

12.3 动态切片

静态切片只能在某些角度给出期望的体渲染效果,这是因为OpenGL的混合效果是依赖于顶点的渲染顺序的。如果能够在摄像机改变位置(模型旋转)时,实时动态地调整顶点,使得这些顶点构成的切片始终能够按照朝向摄像机且从远到近的顺序渲染,那么就可以得到一个无死角的体渲染方法。这就是动态切片的关键点。

最理想的顶点调整方式,是让顶点构成的各个切片始终与摄像机的Near(或Far)裁剪面保持平行,如图12-4所示。

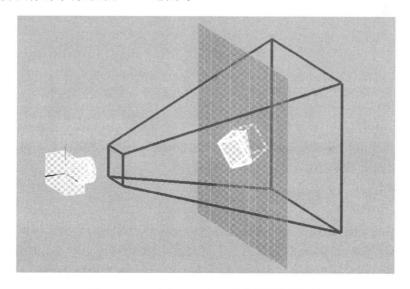

图12-4 一个与Near/Far裁剪面平行的切片

12.3.1 切割立方体

只要找到切片与立方体的12条边的交点,OpenGL就可以用GL_TRIANGLE_FAN、GL_TRIANGLES或GL_POLYGON等方式渲染出此切片。本章使用GL_TRIANGLES的方式,因为这样就可以配合glDrawArrays(..)渲染命令,从而节省了一个IndexBuffer对象。

从数学角度看,切片与立方体可能只有1个交点。但计算机程序使用的浮点数精度是有限的，OpenGL 也只会以相同的方式（如 GL_TRIANGLES）渲染切片。因此，从程序的角度看，只有 1 个交点的情形可以视为有三个位置相同的交点。由于立方体是极度对称的，这里只须考虑摄像机朝向 1 个顶点时的情形即可。切片可能出现的情形如图 12-5 所示。

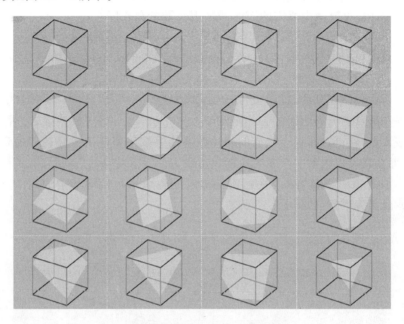

图 12-5　切片可能出现的情形

读者可以在本书网络资料的 c12d02_SlicingSituations 项目，通过调节左侧的 trackbar 来观察不同的切片是如何形成的。

12.3.2　交点的顺序

从数学角度看，求取一个平面与立方体的交点是个简单的解析几何问题。但从 OpenGL 编程角度来看，还要解决另一个问题，即找到这些交点的某个顺序，使得这些交点能顺次围成一个凸多边形。如果顺序不对，就可能出现如图 12-6 所示的畸形切片。

本例使用 GL_TRIANGLES 的方式渲染切片，如果交点的顺序不对，就会产生如图 12-6 所示的不完整的切片，进而影响最终的渲染效果。

一种方法是通过观察切片可能切出的交点在各个边上的顺序，找到一个天然的排列好的交点顺序。在切片从远到近地切割立方体的过程中，可以发现有些边上的交点始终可以排列在另一些边上的交点之前。例如，当切片切割立方体得到 6 个交点时，可以按图 12-7 所示的边的顺序依次求取交点。

将摄像机在 World Space 里的 Front 方向上的单位向量记作 ViewDirection，显然每个切片都垂直于此方向。当摄像机的 Front 方向最接近顶点 s 到顶点 e 所在的

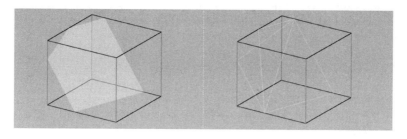

图 12-6 （左）正确的切片　（右）没有覆盖完全的切片的轮廓

直线时，按照图 12-7 所示的边的顺序依次求取交点，可以发现交点会在边 1、2、5、6、9、10 上依次产生。按照这样的顺序，交点恰好可以围成一个凸多边形。此时，只需按照{1 2 5}{1 5 6}{1 6 9}{1 9 10}的索引顺序，将这些交点的位置属性写入 VertexBuffer，即可描述这一切片。而且，无论切片切割立方体的情形如何，都可以按照此图所示的顺序得到天然的排列好的交点。

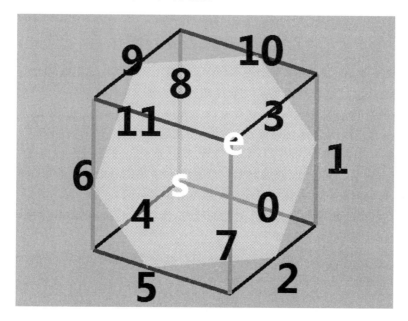

图 12-7　求取交点的特殊顺序

12.4　Raycasting

Raycasting 是一种高效的体渲染算法，可用于交互式的医学成像、科学数据显示等领域。与切片法相同的是，raycasting 也用一个 3D Texture 对象存储体数据。其核心思想：从画布上的一个像素点出发，在垂直于画布的方向上，一条射线穿过描述

3D Texture 对象的立方体,每隔一小段距离采样一次,累加起来就是此像素点应有的颜色,如图 12-8 所示。

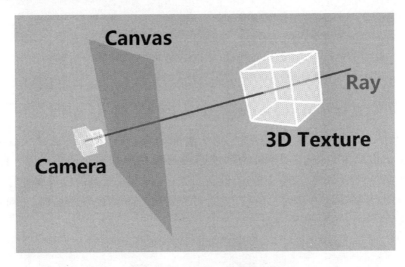

图 12-8 Raycast 原理示意图

从射线接触 3D Texture 对象的第一个点开始(起始点),到射线离开 3D Texture 对象的位置(终点),这段距离就是要采样的范围。那么如何获取终点的位置呢?方法:渲染一个恰好包围 3D Texture 对象的立方体,且只渲染此立方体的背面(用 glCullFace(..);),因为背面就是终点;另外,要把这个渲染结果暂存起来。这就需要一个与画布大小相同的 2D Texture 对象来记录终点的位置,即需要一个新的 Framebuffer 对象。相应地,获取起始点位置的方法就是渲染一个恰好包围 3D 纹理的立方体,且只渲染此立方体的正面(用 glCullFace(..);)。

由于要为每个像素点分别计算累加的颜色,所以 Raycast 的核心算法是在 Fragment Shader 中实现的,代码如下:

```glsl
// fragment shader
#version 150

in vec3 passEntryPoint;

uniform sampler1D texTransfer;   // mapping value to color
uniform sampler2D texExitPoint;  // 2D texture for back face of cube
uniform sampler3D texVolume;     // 3D texture for volume data

uniform float stepLength = 0.001f;
uniform vec2 canvasSize;          // glViewport(..)
uniform vec4 backgroundColor;// value in glClearColor(value)
uniform int cycle = 1600;  // more cycle, more workload
```

```glsl
out vec4 outColor;

void main()
{
    vec3 exitPoint = texture(texExitPoint, gl_FragCoord.st / canvasSize).xyz;

    // if this is background, no raycasting needed
    if (passEntryPoint == exitPoint) { discard; }

    vec3 direction = exitPoint - passEntryPoint;
    float directionLength = length(direction); // the length from front to back
    vec3 deltaDirection = direction / directionLength * stepLength;

    vec3 voxelCoord = passEntryPoint;
    vec3 colorAccumulator = vec3(0.0);// The dest color
    float currentAlpha = 0.0f;
    float currentLength = 0.0;

    for(int i = 0; i < cycle; i++)
    {
        float intensity = texture(texVolume, voxelCoord).x;
        vec4 color = texture(texTransfer, intensity); // The src color
        // front-to-back integration.
        if (color.a > 0.0) {
            currentColor += (1.0 - currentAlpha) * color.a * color.rgb;
            currentAlpha += (1.0 - currentAlpha) * color.a;
        }
        voxelCoord += deltaDirection;
        currentLength += stepLength;
        if (currentLength >= directionLength)
        {
            currentColor = currentColor * currentAlpha
                + (1 - currentAlpha) * backgroundColor.rgb;
            break; // terminate if the ray is outside the volume
        }
        else if (currentAlpha > 1.0)
        {
            currentAlpha = 1.0;
            break; // terminate if opacity > 1
        }
    }

    outColor = vec4(currentColor, currentAlpha);
}
```

此算法按照从近到远的顺序采样,并累加得到一个像素点所在位置的最终颜色。如果设置的背景色与画布的背景色相同,那么体渲染的结果会完全融入背景色中;否则,体渲染的结果会呈现出一个明显的立方体轮廓,如图 12-9 所示。

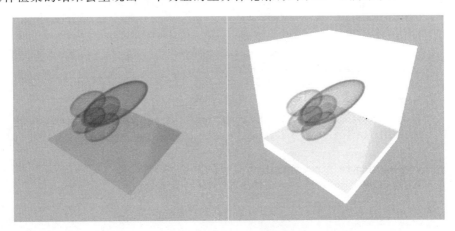

图 12-9　(左)融入背景色　(右)呈现立方体轮廓

实现 Raycast 的一个难点在于,要分别初始化 1 个 1D Texture 对象、1 个 2D Texture 对象、1 个 3D Texture 对象和 1 个 Framebuffer 对象。需要注意的是,在某些硬件上将不同类型的 Texture 对象绑定到同一个 Texture Unit 上可能会失败。因此,最好将各个 Texture 对象分别绑定到不同的 Texture Unit。初始化的代码如下:

```
public partial class RaycastNode
{
    private Texture texTransfer;
    private Texture texBackface;
    private int width;
    private int height;
    private Texture texVolume;
    private Framebuffer framebuffer;

    protected override void DoInitialize()
    {
        base.DoInitialize();

        this.texTransfer = InitTransferTexture();
        byte[] volumeData = GetVolumeData();
        this.texVolume = InitVolumeTexture(volumeData, 256, 256, 256);
        {
            // setting uniforms
```

```csharp
            RenderMethod method = this.RenderUnit.Methods[1];
            ShaderProgram program = method.Program;
            program.SetUniform("texTransfer", this.texTransfer);
            program.SetUniform("texVolume", this.texVolume);
        }
    }

    // reinitialize when canvas is resized
    private void Resize(int width, int height)
    {
        if (this.texBackface != null) { this.texBackface.Dispose(); }
        if (this.framebuffer != null) { this.framebuffer.Dispose(); }

        this.texBackface = InitBackfaceTexture(width, height);
        this.framebuffer = InitFramebuffer(width, height, this.texBackface);
        {
            RenderMethod method = this.RenderUnit.Methods[1];
            ShaderProgram program = method.Program;
            program.SetUniform("canvasSize", new vec2(width, height));
            program.SetUniform("texExitPoint", this.texBackface);
        }
    }

    private Framebuffer InitFramebuffer(int width, int height, Texture texture)
    {
        var framebuffer = new Framebuffer(width, height);
        framebuffer.Bind();
        framebuffer.Attach(texture, 0u); // location is 0u
        {
            var depthBuffer = new Renderbuffer(width, height, GL_DEPTH_COMPONENT24);
            framebuffer.Attach(depthBuffer, AttachmentLocation.Depth);
        }
        framebuffer.CheckCompleteness();
        framebuffer.Unbind();

        return framebuffer;
    }

    private byte[] GetVolumeData(string filename)
    {
```

```csharp
            using (var fs = new FileStream(filename, FileMode.Open, FileAccess.Read))
            using (var br = new BinaryReader(fs))
            {
                int unReadCount = (int)fs.Length;
                data = new byte[unReadCount];
                br.Read(data, 0, unReadCount);
            }

            return data;
        }

        private Texture InitVolumeTexture(byte[] data, int width, int height, int depth)
        {
            var storage = new TexImage3D(GL_Texture_3D, GL_RED, width, height, depth, GL_RED, GL_UNSIGNED_BYTE, new ArrayDataProvider<byte>(data));
            var texture = new Texture(storage, new MipmapBuilder(),
                new TexParameteri(GL_TEXTURE_WRAP_R, GL_CLAMP),
                new TexParameteri(GL_TEXTURE_WRAP_S, GL_CLAMP),
                new TexParameteri(GL_TEXTURE_WRAP_T, GL_CLAMP),
                new TexParameteri(GL_TEXTURE_MIN_FILTER, GL_LINEAR_MIPMAP_LINEAR),
                new TexParameteri(GL_TEXTURE_MAG_FILTER, GL_LINEAR),
                new TexParameteri(GL_TEXTURE_BASE_LEVEL, 0),
                new TexParameteri(GL_TEXTURE_MAX_LEVEL, 4));
            texture.Initialize();
            texture.TextureUnitIndex = 2;

            return texture;
        }

        private Texture InitBackfaceTexture(int width, int height)
        {
            var storage = new TexImage2D(GL_Texture_2D, GL_RGBA16F, width, height, GL_RGBA, GL_FLOAT);
            var texture = new Texture(storage,
                new TexParameteri(GL_TEXTURE_WRAP_R, GL_REPEAT),
                new TexParameteri(GL_TEXTURE_WRAP_S, GL_REPEAT),
                new TexParameteri(GL_TEXTURE_WRAP_T, GL_REPEAT),
                new TexParameteri(GL_TEXTURE_MIN_FILTER, GL_NEAREST),
                new TexParameteri(GL_TEXTURE_MAG_FILTER, GL_NEAREST));
            texture.Initialize();
            texture.TextureUnitIndex = 1;
```

```
        return texture;
    }


    private Texture InitTransferTexture(string filename)
    {
        byte[] tff;
        using (var fs = new FileStream(filename, FileMode.Open, FileAccess.Read))
        using (var br = new BinaryReader(fs))
        {
            tff = br.ReadBytes((int)fs.Length);
        }


        const int width = 256;
        var storage = new TexImage1D(GL_RGBA8, width, GL_RGBA, GL_UNSIGNED_BYTE, new
                    ArrayDataProvider<byte>(tff));
        var texture = new Texture(storage,
            new TexParameteri(GL_TEXTURE_WRAP_R, GL_REPEAT),
            new TexParameteri(GL_TEXTURE_WRAP_S, GL_REPEAT),
            new TexParameteri(GL_TEXTURE_WRAP_T, GL_REPEAT),
            new TexParameteri(GL_TEXTURE_MIN_FILTER, GL_NEAREST),
            new TexParameteri(GL_TEXTURE_MAG_FILTER, GL_NEAREST));
        texture.Initialize();
        texture.TextureUnitIndex = 0;


        return texture;
    }
}
```

当然，也可以先渲染起始点，然后在找到终点的时候计算各个像素点的颜色值。

12.5 总　结

本章介绍了实现体渲染的 2 种方法。

① 切片成功的关键是什么？

切片与立方体交点的排列顺序。

② Raycast 算法的核心是什么？

对每个像素累加颜色。

12.6 问 题

带着问题实践是学习 OpenGL 最快的方式。这里给读者提出几个问题,作为抛砖引玉之用。

1. 请在本书网络资料的 c12d00_StaticSlices 项目中,通过修改 Shader 的方式让渲染结果呈现出如图 12-3 所示的样式。

2. 请针对动态切片切割立方体的各种可能情形,分别检验本章给出的边的顺序,是否能够得到天然的排列好的交点?

3. 自定义 Raycast 中的 1D Texture 对象的内容并观察效果,思考 1D Texture 在此的意义是什么?

4. 修改 Raycasting 算法,使之以从远到近的方式累加颜色。

第 13 章

半透明渲染

学完本章之后,读者将能:
- 使用跳跃着色法实现简单的半透明渲染;
- 用 OIT 算法实现半透明渲染;
- 用 Front to Back Peeling 算法实现半透明渲染。

OpenGL 渲染半透明物体是一个比较复杂和困难的问题。本章介绍几种适用于不同情形的半透明渲染方法。

13.1 跳跃着色法

如果只渲染坐标值为奇数(或偶数)的 Fragment,就可以形成一种半隐半现的效果。这是一种对半透明渲染的近似方法。代码如下:

```
// fragment shader
#version 150 core

// ...

out vec4 outColor;

void main()
{
    // skip half of all fragments
    if (int(gl_FragCoord.x + gl_FragCoord.y) % 2 == 1) discard;

    outColor = ...
}
```

读者可以在本书网络资料中找到 c13d00_SkipFragCoord 项目,观察将 Cube 设置为"半透明"的效果,如图 13-1 所示。

如果放大画布,就可以看到渲染 Cube 的 Fragment 被以跳跃式的方式忽略了,如图 13-2 所示。

图 13-1　半透明的 Cube 包裹部分 Teapot

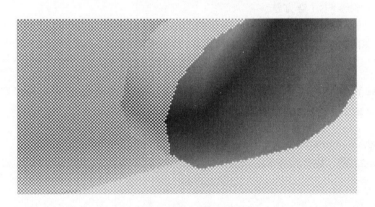

图 13-2　放大观察"半透明"的 Cube

由于 Fragment 排列紧密,且 Fragment Shader 均匀地忽略了一半 Fragment,这就实现了视觉上的半透明效果。这一方法十分简单,但仅适用于半透明的物体互不遮挡的简单场景。本书中,很多示例都采用这一方法。

13.2　与顺序无关的半透明渲染

与顺序无关的透明算法(OIT,Order-Independent-Transparency)是一个效果很好的半透明渲染算法。为了解决在通常的混合过程中可能出现的"后渲染的 Fragment 比先渲染的 Fragment 更远"的情况,OIT 算法将一个像素点上的所有 Fragment 存储到一个链表上,对链表排序,最后对排序好的 Fragment 依次渲染,从而得到此像素上期望的半透明效果。也就是说,OIT 的核心思想是分两遍渲染:第一遍,把像素上的所有 Fragment 都存到一个链表里;第二遍,根据每个 Fragment 的深度值对链表进行排序,然后进行混合操作,这就解决了混合顺序的问题。本算法是根据一位 AMD 的工程师的设计而来。

13.2.1　Width * Height 个链表

为了对宽度为 Width、高度为 Height 的画布实施 OIT 渲染,必须为此画布上的每个像素分别设置一个链表,用于存储投影到此像素上的各个 Fragment,即一共需要"Width * Height"个链表,这个链表要由 Fragment Shader 来操作(增删节点、排序等)。Shader 本身没有可直接使用的链表数据结构,因此要用一种更原始的方式实现链表操作,即用一个足够大的 Texture 来保存所有的链表节点。这个 Texture 的类型是 GL_TEXTURE_BUFFER,其内容由一个类型为 GL_TEXTURE_BUFFER 的 Buffer 进行初始化,这个 Texture 将存储"Width * Height"个链表。设想 OIT 执行第一遍渲染时,这个 Texture 的第二个位置上可能是第一个像素的第二个 Fragment 的节点,也可能是第二个像素的第一个 Fragment 的节点。也就是说,单凭此 Texture 是无法迅速找到各个像素上的链表的,这就意味着还需要一个数组来记录"Width * Height"个链表的头节点的位置。这个数组可以用一个尺寸为"Width * Height"的 2D Texture 对象实现。

由记录头节点位置的 2D Texture 和记录各个节点的 Texture 构成的 OIT 算法的核心数据结构如图 13-3 所示。

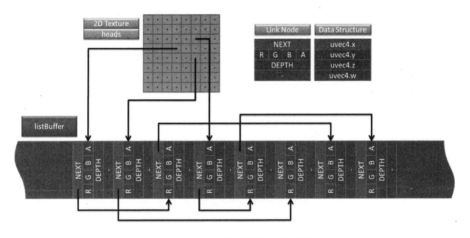

图 13-3　OIT 算法的核心数据结构

图 13-3 中,每个节点由一个 uvec4 结构体表示。uvec4 的第一个字段(uvec4.x)用于存储 NEXT 指针;uvec4 的第二个字段(uvec4.y)用于存储 Fragment 的 RGBA 颜色;uvec4 的第三个字段(uvec4.z)用于存储 Fragment 的深度值,稍后它将用于对 Fragment 排序;uvec4 的第四个字段(uvec4.w)用于扩展 OIT 的功能,暂时可以忽略。

从GL_TEXTURE_BUFFER类型的Buffer生成的Texture对象,此时被视为一块连续的内存。从内存的起始位置开始紧密地存储着"Width * Height"个链表的所有节点,每个节点的NEXT指针记录的是下一个节点在此Texture中的位置。为叙述方便,本章将此Texture称为lstBuffer。注意,这里的2D Texture不像一般的Texture对象那样用于给模型贴图,而是用作记录头节点的位置。为叙述方便,本章将此Texture称为heads,这说明Texture对象是一种十分灵活的数据结构,其用法是极其多变的。

13.2.2 构建链表

第一遍渲染时,Fragment Shader负责构建自己所在像素上的链表,代码如下:

```glsl
//vertex shader for building link lists
#version 330

in vec3 inPosition;

uniform mat4 mvpMatrix;

out vec4 passColor;

void main()
{
    gl_Position = mvpMatrix * vec4(inPosition, 1.0f);

    passColor = ... // get color of this fragment somehow
}

//----------------------------------------------------
//fragment shader for building link lists
#version 420 core

layout (early_fragment_tests) in; // enable early fragment testing

// atom operation lock
uniform atomic_uint lstCounter;
// The per-pixel image containing the head pointers
layout (binding = 0, r32ui) uniform uimage2D heads;
```

```glsl
// Buffer containing linked lists of fragments
layout (binding = 1, rgba32ui) uniform uimageBuffer lstBuffer;

in vec4 passColor; // color of this fragment

out vec4 color; // this is not used

void main()
{
    uint index = atomicCounterIncrement(lstCounter); // position of next link node

    uint preHead = imageAtomicExchange(heads, ivec2(gl_FragCoord.xy), index);

    uvec4 item;
    item.x = preHead; // the 'NEXT' pointer
    item.y = packUnorm4x8(passColor); // compress RGBA color to an uint
    item.z = floatBitsToUint(gl_FragCoord.z); // depth value

    imageStore(lstBuffer, int(index), item); // write new link node to lstBuffer

    discard; // no need to render to canvas.
}
```

首先，用 atomicCounterIncrement(lstCounter)函数返回的 index 是 lstBuffer 中下一个尚未使用的节点的位置。lstCounter 是 atomic_uint 类型，内置函数 atomicCounterIncrement(..)的作用是将指定的变量增加 1，并且返回增加 1 之前的数值。它可以保证在增加 1 的过程中不被高度并行化执行的 Shader 扰乱，即实现一个增加 1 的原子操作。显然，Fragment Shader 每次执行都会调用一次 atomicCounterIncrement(lstCounter)函数。在它被调用 N 次（$N \geqslant 1$）后，lstCounter 保存的值为 N，函数返回的值则为（$N-1$）。在 atomicCounterIncrement(lstCounter)函数执行后的内存状态如图 13-4 所示。

然后，用 imageAtomicExchange(..)函数将 heads 中当前像素位置上的头节点指针指向 index 的位置，并获取在此改变之前的头节点位置 preHead。

heads 的类型是 uimage2D，这种特别的图像类型可以表示一个 Buffer 对象或者一个 Texture 的某一 level，并且支持在 Shader 中进行读写操作。如需在 Shader 中使用 uimage2D 类型的变量，那么在 OpenGL 客户端就应将某个 Buffer 或 Texture 绑定到一个图像单元(Image Unit)。图像单元与纹理单元十分相似，区别在于纹理单元用于绑定 Texture(Shader 中对应于 sampler2D 等)，图像单元用于绑定 Buffer 或 Texture 的某一 level。实现图像单元绑定的 OpenGL 函数为：

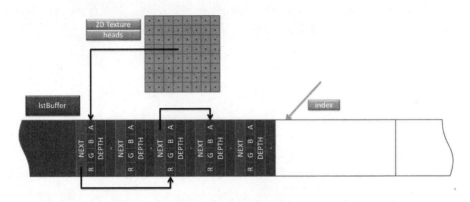

图 13-4 找到下一个尚未使用的节点的位置 index

```
// bind specified level of texture with textureId to specified imageUnit
void glBindImageTexture(uint imageUnit, uint textureId, int level, bool layered, int
    layer, uint access, uint format);
```

lstBuffer 的类型是 uimageBuffer,这也是一种图像类型,并且支持在 Shader 中进行读写操作。如需在 Shader 中使用 uimageBuffer 类型的变量,那么在 OpenGL 客户端应依据某个 Buffer 的内容和长度来创建 Texture,并将此 Texture 绑定到一个图像单元(Image Unit)。根据 Buffer 创建 Texture 需要用到的核心 OpenGL 函数为:

```
//setup texture's content according to specified buffer with bufferId
void glTexBuffer(uint target, uint format, uint bufferId);
```

在 imageAtomicExchange(..)函数执行后的内存状态如图 13-5 所示。

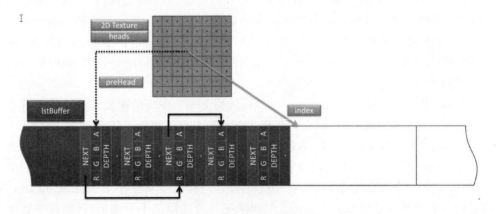

图 13-5 修改头节点指针

然后,创建新节点。新节点的 xyz 字段分别存储了旧的头节点位置 preHead、压缩格式的 Fragment 的 RGBA 颜色值和深度值,如图 13-6 所示。

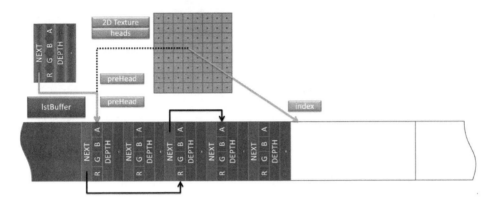

图 13-6 创建新节点

注意,压缩存储的颜色值 RGBA 是用一个 uint 值表示的,GLSL 内置函数 packUnorm4x8(..)可以在客户端实现如下:

```
// C# code
public static partial class glm
{
    // pack vec4 to uint.
    public static uint packUnorm4x8(this vec4 color)
    {
        uint x = (uint)(Math.Round(clamp(color.x, 0.0f, 1.0f) * 255.0));
        uint y = (uint)(Math.Round(clamp(color.y, 0.0f, 1.0f) * 255.0));
        uint z = (uint)(Math.Round(clamp(color.z, 0.0f, 1.0f) * 255.0));
        uint w = (uint)(Math.Round(clamp(color.w, 0.0f, 1.0f) * 255.0));
        uint result = (w << 24) | (z << 16) | (y << 8) | (x);
        return result;
    }

    // clamp value between min and max
    private static float clamp(float value, float min, float max)
    {
        if (value < min) { return min; }
        if (max < value) { return max; }
        return value;
    }
}
```

相应地,将 uint 值表示的压缩的 RGBA 转换为 vec4 类型的 unpackUnorm4x8 (..) 实现如下:

```
// C# code
public static partial class glm
{
    // unpack uint to vec4.
    public static vec4 unpackUnorm4x8(this uint p)
    {
        const uint mask8 = (1 << 8) - 1; // 0xFF, 255u
        uint x = (p >> 0) & mask8;
        uint y = (p >> 8) & mask8;
        uint z = (p >> 16) & mask8;
        uint w = (p >> 24) & mask8;

        var result = new vec4(x / 255.0f, y / 255.0f, z / 255.0f, w / 255.0f);

        return result;
    }
}
```

然后,用 imageStore(lstBuffer, int(index), item) 函数将新节点 item 写入 lstBuffer 中下一个尚未使用的节点的位置,即 index 的位置,如图 13-7 所示。

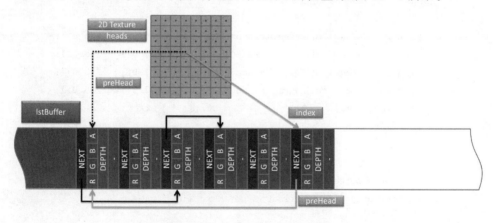

图 13-7　新节点写入 lstBuffer

最后,用 discard 关键字通知 OpenGL,此 Fragment 不需要被渲染到画布上。经过第一遍渲染后,每个像素上都有一个链表与之对应。链表的每个节点都存储着一个 Fragment 的颜色值和深度值。

13.2.3 排序和混合

第二遍渲染时，Fragment Shader 负责对自己所在像素上的链表进行排序并输出混合颜色，代码如下：

```
//vertex shader for sorting and blending color
#version 420 core

in vec3 inPosition;
uniform mat4 mvpMatrix;

voidmain()
{
    gl_Position = mvpMatrix * vec4(inPosition);
}

// -----------------------------------------------------------
//fragment shader for sorting and blending color
#version 420 core

// The per-pixel image containing the head pointers
layout (binding = 0, r32ui) uniform uimage2Dheads;
// Buffer containing linked lists of fragments
layout (binding = 1, rgba32ui) uniform uimageBuffer lstBuffer;

// This is the output color
out vec4 color;

// This is the maximum number of overlapping fragments allowed
const int maxFragment 40

// Temporary array used for sorting fragments
uvec4 lstFragment[maxFragment];

void main()
{
    uintfragmentCount = 0;
    // pointer of head node
    uint currentIndex = imageLoad(heads, ivec2(gl_FragCoord).xy).x;
    // find all fragments for current pixel
```

```glsl
    while (currentIndex != 0 && fragmentCount < maxFragment)
    {
        uvec4 fragment = imageLoad(lstBuffer, int(currentIndex));
        lstFragment[fragmentCount] = fragment;
        currentIndex = fragment.x; // the 'NEXT' field.
        fragmentCount ++ ;
    }
    // sort fragments by depth
    for (uint i = 0; i < fragmentCount - 1; i ++ )
    {
        uint p = i;
        uint depth1 = (lstFragment[p].z);
        for (uint j = i + 1; j < fragmentCount; j ++ )
        {
            uint depth2 = (lstFragment[j].z);
            if (depth1 < depth2)
            {
                p = j; depth1 = depth2;
            }
        }
        if (p != i)
        {
            uvec4 tmp = lstFragment[p];
            lstFragment[p] = lstFragment[i];
            lstFragment[i] = tmp;
        }
    }
    // blend
    vec4 dest = vec4(0.0);
    for (uint i = 0; i < fragmentCount; i ++ )
    {
        vec4 src = unpackUnorm4x8(lstFragment[i].y);
        // same with glBlend(GL_SRC_ALPHA, GL_ONE_MINUS_SRC_ALPHA)
        dest = src * src.a + dest * (1.0f - src.a);
    }
    color = dest;
}
```

首先,从lstBuffer中找到当前像素上所有的Fragment,并保存到一个临时数组uvec4 lstFragment[maxFragment]中。可以发现,如果一个像素上的Fragment太多,OIT算法会自动忽略部分Fragment。然后,根据Fragment的深度值对其排序。

这是一个对定长数组排序的任务，这里使用了选择排序算法。最后，在 Shader 中模拟混合操作，即按照从远到近的顺序，用与 glBlend(GL_SRC_ALPHA, GL_ONE_MINUS_SRC_ALPHA) 相同的方式计算混合后的颜色值。

13.2.4 其 他

Modern OpenGL 是用 GLSL Shader 来实施渲染的。Shader 是一段程序，程序的创造力是无穷无尽的，开发者可以以任何方式使用 OpenGL 提供的资源（Texture、Buffer 等）。唯一固定不变的，就是 OpenGL 的渲染管道（pipeline）。记住 pipeline 的工作流程，认识 OpenGL 的各种资源，发挥想象力，就可以创造出绚丽多彩的三维场景。

本章以及本书多次使用的 2 遍渲染或多遍渲染的成功实验，说明了 OpenGL 只负责渲染，至于渲染之后是画到画布还是其他地方，都是可以控制的。甚至，渲染的目的原本就不必是为了画图，这就是以 Compute Shader 为代表的 GPU 通用计算。

13.3 Front to Back Peeling

13.3.1 过 程

Front-to-Back-Peeling 用另一种方式解决了 Fragment 渲染顺序的问题。它在一个循环中多次渲染同一模型，每次只渲染每个像素上最靠近摄像机的 Fragment，然后用 discard 关键字屏蔽这些 Fragment。这样，下次循环时就会只渲染每个像素上第二靠近摄像机的 Fragment，如此循环直至处理完所有的 Fragment。

如何逐层地屏蔽 Fragment 呢？在 Pipeline 流程中，深度测试只会保留最接近摄像机的那层 Fragment。这恰是第一次循环过程要渲染的内容。另外，如果用一个 Texture 记录这些 Fragment 的深度值（像在 Shadow Mapping 中那样），那么下次循环时，就可以在 Shader 中将深度值小于等于 Texture 中记录的 Fragment 忽略掉，从而渲染了第二靠近摄像机的 Fragment。为叙述方便，本小节将此 Texture 称为 blenderColor。

这样，经过若干次重复渲染模型，就可以将整个模型一层一层地渲染出来。这一渲染过程中，每个像素上的 Fragment 都是依次从近到远出现的。只要用 glBlendFuncSeparate(GL_DST_ALPHA, GL_ONE, GL_ZERO, GL_ONE_MINUS_SRC_ALPHA) 进行混合，就可以得到期望的半透明效果。

OpenGL 的函数 glBlendFuncSeparate(..) 与 glBlendFunc(..) 类似，区别在于它将对 RGB 分量的混合参数和对 Alpha 分量的混合参数分开指定，从而具有更多的灵活性。

Front-to-Back-Peeling 的大致过程如下：

首先，要进行一次初始化操作，目的是让 blenderColor 记录最靠近摄像机的那层 Fragment 的深度值。根据上一次渲染时的深度信息，得到第二层 Fragment，并存入一个 Texture，这是切割（peel）的过程；然后让第二层 Fragment 与之前的渲染结果混合（blend）。重复这一操作，直至所有的 Fragment 都处理完毕。最后，将背景色（background）混合到当前的渲染结果，遂成最终结果（final）。

以 27 个均匀排列的透明的半立方体为例，整个过程如图 13-8 所示。

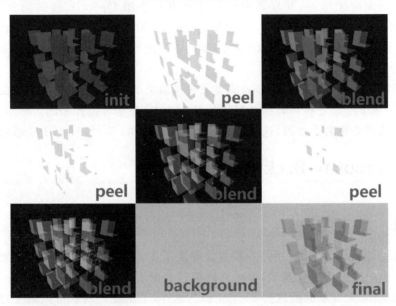

图 13-8　Front-to-Back-Peeling 过程示意图

Front-to-Back-Peeling 的中间过程需要暂存一些内容。具体地说，需要为此算法创建 3 个自定义的 Framebuffer，如图 13-9 所示。

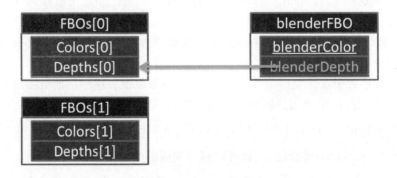

图 13-9　用于暂存数据的自定义 Framebuffer

如图 13-9 所示，首先创建一个 Framebuffer 数组 FBOs[2]，用于交替实施 Peel

和 Blend 操作。另外再创建一个 Framebuffer 对象 blenderFBO,用于初始化等操作。注意,blenderFBO 的深度缓存用的是 FBOs[0]的深度缓存对象。

初始化过程中,启用 blenderFBO,这样在第一遍渲染完成后,blenderColor 记录了最新的渲染结果,而 FBOs[0]的深度缓存 Depths[0]就记录了此时最靠近摄像机的那层 Fragment 的深度信息,如图 13-10 所示。

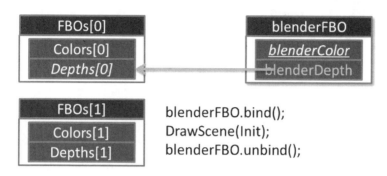

图 13-10 初始化时填充了 Depths[0]

在第二遍渲染时启用 FBOs[1],根据 Depths[0]记录的深度信息,可以用 discard 关键字忽略最靠近摄像机的 Fragment,从而使得第二靠近摄像机的 Fragment 被渲染到 Colors[1]上。此时的 Depths[1]记录了最新的深度信息,如图 13-11 所示。

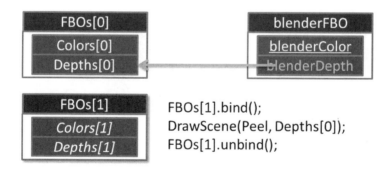

图 13-11 执行一次 Peel 操作

在第三遍渲染时,启用 blenderFBO,将记录着最新的 Fragment 层的 Colors[1]的内容用混合的方式写入当前的渲染结果。注意,这次不是渲染场景,而是渲染一个恰好覆盖画布的四边形(QUAD),这使得 blenderColor 再次获得了最新的渲染结果,如图 13-12 所示。

不断地重复上述 2 个步骤,直至所有 Fragment 都已被逐层地切割并混合。注意,每次重复上述 2 个步骤时,FBOs[2]数组的 2 个元素都交替负责对方的工作。也就是说,第四遍和第五遍渲染时的自定义 Framebuffer 状态如图 13-13 所示。

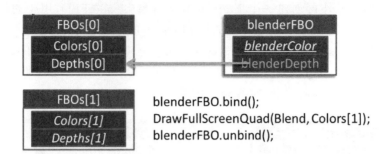

图 13-12　执行一次 Blend

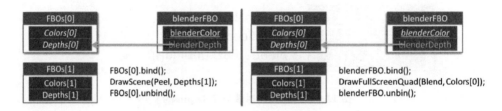

图 13-13　再次执行 Peel 和 Blend

最后,将背景色用混合的方式写入当前的渲染结果,就得到了最终的渲染结果。也就是说,要将 blenderColor 和背景色进行混合,并写入默认的 Framebuffer,如图 13-14 所示。

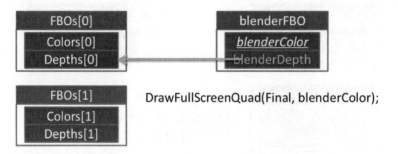

图 13-14　将 background 混合写入渲染结果

13.3.2　查询对象

上文提到,当所有的 Fragment 都处理完时,Peel 和 Blend 的循环操作就结束了。那么如何判断所有的 Fragment 是否已经处理完?

使用查询对象(Query Object),可以方便地解决这一问题。

查询对象与缓存对象、纹理对象一样,是一种 OpenGL 对象。只需在渲染前开启它,在渲染后结束它,它就可以告诉开发者渲染过程中出现了多少个 Fragment。

如果出现的 Fragment 数量为 0,就说明 Peel 和 Blend 的循环操作应当结束了。创建查询对象与创建其他 OpenGL 对象类似,代码如下:

```csharp
void HowToCreateQuery()
{
    Query query = new Query();
}

public partial class Query : IDisposable
{
    protected uint[] ids = new uint[1];

    public uint Id { get { return this.ids[0]; } }

    // Create a query object
    public Query()
    {
        glGenQueries(1, this.ids);
    }

    // Begin query
    public void BeginQuery(QueryTarget target)
    {
        glBeginQuery((uint)target, this.Id);
    }

    // End query
    public void EndQuery(QueryTarget target)
    {
        glEndQuery((uint)target);
    }

    // How many fragments have been rendered
    public int SampleCount()
    {
        var result = new int[1];
        glGetQueryObjectiv(this.Id,GL_QUERY_RESULT, result);

        return result[0];
    }
}
```

使用查询对象判断出现了多少个Fragment从而终止渲染的代码如下：

```
var query = new Query();
int sampleCount;
do
{
    query.BeginQuery(QueryTarget.SamplesPassed); // begin monitoring fragments
    DrawScene();
    query.EndQuery(QueryTarget.SamplesPassed); // stop monitoring fragments

    sampleCount = query.SampleCount(); // How many fragments rendered
} while (sampleCount > 0)
```

读者可在本书网络资料的c13d01_QueryObject项目中找到介绍Query对象的完整代码。查询对象还有多种功能和用法，暂不详述。

13.4 总　结

本章介绍了几种实现半透明渲染的方法。
① 跳跃着色法的适用范围是什么？
适用于半透明的物体互不遮挡的简单场景。
② OIT算法的核心是什么？
对每个像素上的Fragment，先排序后混合。
③ Front to Back Peeling的核心是什么？
多次渲染模型，每次只取最靠近摄像机的Fragment。

13.5 问　题

带着问题实践是学习OpenGL最快的方式。这里给读者提出几个问题，作为抛砖引玉之用。

1. 请在本书网络资料中查找任意一个使用了跳跃着色法的项目，尝试关闭算法，观察效果。

2. 请用本书网络资料中的Demos\OrderIndependentTransparency项目观察几个模型，并用普通的渲染方法观察混合的结果，相互对比。

3. 尝试使用不同的Peel顺序和Blend方式实现Peeling算法。

4. 在Peeling算法中使用了3个自定义的Framebuffer对象，尝试减少需要的自定义Framebuffer对象。

第 14 章

Transform Feedback Object

学完本章之后,读者将能:
- 理解有 Transform Feedback 参与的 Pipeline;
- 认识和使用 Transform Feedback Object;
- 使用 Transform Feedback 功能实现简单的粒子系统。

Transform Feedback 是一项更新顶点缓存(VertexBuffer)的技术。使用 glMapBuffer(..)可以读写顶点缓存,但是这需要在 CPU 和 GPU 之间传输数据,其效率较低。而使用 Transform Feedback 技术可以直接在 GPU 端并行地读写顶点缓存。

14.1 Transform Feedback 如何工作

本小节通过一个简单的示例介绍 Transform Feedback 的工作方式。读者可在本书网络资料的 c14d00_HowTransformFeedbackWorks 项目中找到完整代码。本示例用 Transform Feedback 为一个数组中的每个浮点数计算其开方值。计算过程在 Vertex Shader 中实现,代码如下:

```
// vertex shader
#version 150 core

in float inValue;
out float outValue;

void main()
{
    outValue = sqrt(inValue);
}
```

输入变量 inValue 的数据由一个 VertexBuffer 提供,这可以通过绑定 ShaderProgram 中的属性位置与此 VertexBuffer 来实现。在本书的各个示例中都有这一知识点的应用,不再赘述。

通常,输出变量outValue的数据会被下一阶段的Shader(例如Geometry Shader或Fragment Shader)接收。但本例中它由另一个VertexBuffer接收,这需要使用Transform Feedback来实现。与输入变量类似,Transform Feedback通过标记outValue变量的名称来寻找它,然后通过绑定此标记与另一个VertexBuffer来实现。此时,这个VertexBuffer就被称为Transform Feedback Buffer。代码如下:

```csharp
// create a transform feedback calculator instance
public TransformFeedbackCalculator(int maxItemCount)
{
    this.MaxItemCount = maxItemCount;

    var program = new ShaderProgram();
    uint id = program.ProgramId; // uint id = glCreateProgram()
    var varyings = new string[] { "outValue" };
    // record out variables in vertex shader
    glTransformFeedbackVaryings(id, varyings.Length, varyings, GL_INTERLEAVED_ATTRIBS);

    varvs = new VertexShader(vertexCode);
    vs.Initialize();
    glAttachShader(id, vs.ShaderId);
    glLinkProgram(id);
    program.CheckLinkStatus();

    // input buffer
    VertexBuffer vbo = VertexBuffer.Create(typeof(float), maxItemCount, VBOConfig.Float, BufferUsage.StaticDraw);
    var drawCmd = new DrawArraysCmd(DrawMode.Points, maxItemCount);
    var attribute = new VertexShaderAttribute(vbo, "inValue");
                                                    // connect vbo to "inValue"
    var vao = new VertexArrayObject(drawCmd, program, attribute);
    uint index = 0; // transform feedback binding point index
    // transform feedback buffer object
    VertexBuffer tbo = VertexBuffer.Create(typeof(float), maxItemCount, VBOConfig.Float, BufferUsage.StaticRead);
    // bind tbo to transform feedback binding point index
    glBindBufferBase(GL_TRANSFORM_FEEDBACK_BUFFER, index, tbo.BufferId);
}
```

计算平方根的过程就是"渲染"的过程。类似使用Query Object的方式,需要Transform Feedback发挥作用时,就在调用OpenGL渲染命令前启用,在调用OpenGL渲染命令后关闭。代码如下:

Transform Feedback Object

```
public void Calculate()
{
    glEnable(GL_RASTERIZER_DISCARD);

    this.program.Bind();
    glBeginTransformFeedback(GL_POINTS); // transform feedback begins
    this.vao.Draw(); // glDrawArrays(GL_POINTS, ..)
    glEndTransformFeedback();
                    // transform feedback ends. Now output VertexBuffer is updated
    this.program.Unbind();

    glDisable(GL_RASTERIZER_DISCARD);
}
```

注意，这一渲染过程中，没有 Fragment Shader 参与，也不需要其参与。因此使用 glEnable(GL_RASTERIZER_DISCARD) 关闭了 Fragment Shader 阶段及其后续阶段。

注意，在更新顶点属性的 ShaderProgram 中没有启用 Geometry Shader。如果启用了 Geometry Shader，Transform Feedback 功能会接收到 Geometry Shader 阶段完成后的输出结果。之后的 Fragment Shader 阶段仍然可以用 glEnable(GL_RASTERIZER_DISCARD) 忽略掉。

计算过程（即渲染过程）完成后，平方根结果就保存到了 outValue 对应的 VertexBuffer 中。可以用下述方式直接获取其中的数据：

```
public float[] GetOutput(int count)
{
    var array = new float[count];
    GCHandle pinned = GCHandle.Alloc(array, GCHandleType.Pinned);
    IntPtr header = Marshal.UnsafeAddrOfPinnedArrayElement(array, 0);
    // copy data to array
    Int index = 0; // transform feedback binding point index
    glGetBufferSubData(GL_TRANSFORM_FEEDBACK_BUFFER, index, sizeof(float) * count, header);
    pinned.Free();

    return array;
}
```

也可以用 glMapBuffer(..) 的方式来获取其中的数据：

```
public unsafe float[] GetOutput(int count)
{
    vararray = new float[count];
    VertexBuffer buffer = this.outputBuffer;
    var pointer = (float*)buffer.MapBuffer(MapBufferAccess.ReadOnly);
    for (int i = 0; i < array.Length && i < this.MaxItemCount; i++)
    {
        array[i] = pointer[i];
    }
    buffer.UnmapBuffer();

    returnarray;
}
```

14.2 使用 Transform Feedback Object

本小节使用 Transform Feedback Object 来整理归纳对 Transform Feedback 的应用,使用 Transform Feedback Object 便于实现一些稍微复杂的实际功能。读者可在本书网络资料的 c14d01_TransformFeedbackObject 项目中找到本小节的全部代码。

计算平方根的 Vertex Shader 与上一小节相同。不同的是,绑定 outValue 变量与 VertexBuffer 的功能通过一个 Transform Feedback Object 来实现,代码如下:

```
public TransformFeedbackCalculator(int maxItemCount)
{
    uint index = 0;
    VertexBuffer tbo = VertexBuffer.Create(typeof(float), maxItemCount, VBOConfig.
                       Float, BufferUsage.StaticRead);
    //glBindBufferBase(GL_TRANSFORM_FEEDBACK_BUFFER, index, tbo.BufferId);
    var tfo = new TransformFeedbackObject();
    tfo.BindBuffer(index, tbo); // call glBindBufferBase(..) inside.
    //...
}
```

注意,一个 Transform Feedback Object 有多个绑定点(Binding Point),每个绑定点可以绑定一个不同的 VertexBuffer(或者同一个 VertexBuffer 的不同部分),绑定点由索引编号 index 指定。

计算平方根的过程就是渲染的过程。类似于使用 Query Object 的方式,需要 Transform Feedback Object 发挥作用时,就在调用 OpenGL 渲染命令前启用,在调用 OpenGL 渲染命令后关闭。代码如下:

```
public void Calculate()
{
    ShaderProgram program = this.program;
    TransformFeedbackObject tfo = this.tfo;
    VertexArrayObject vao = this.vao;

    glEnable(GL_RASTERIZER_DISCARD);

    program.Bind();

    tfo.Bind();//glBindTransformFeedback(GL_TRANSFORM_FEEDBACK, tfo.Id)
    tfo.Begin(GL_POINTS);//glBeginTransformFeedback(GL_POINTS)

    vao.Draw(); // render something

    tfo.End();//glEndTransformFeedback()
    tfo.Unbind();//glBindTransformFeedback(GL_TRANSFORM_FEEDBACK, 0)

    program.Unbind();

    glDisable(GL_RASTERIZER_DISCARD);
}
```

在启用 Transform Feedback Object 前要用 Bind() 来指定它为当前的 Transform Feedback Object,这样之后的操作才会针对它进行;对应地,在关闭 Transform Feedback Object 后要取消指定。

在上一小节中,没有 Bind() 和 Unbind() 操作。可以认为是针对默认的 Transform Feedback Object(其编号为 0)进行操作。

14.3 轮流更新

本小节通过一个简单的示例演示如何使用 2 个 Transform Feedback Object 轮流更新 2 个 VertexBuffer 的内容并渲染到画布,应用这一设计可以实现很多有趣的功能。本小节的示例演示了如何实时更新一个三角形的三个顶点的位置。读者可在本书网络资料的 c14d02_DoubleTransformFeedbakObjects 项目中找到本小节示例的全部代码。更新顶点位置的 Shader 代码如下:

```glsl
//updateVert.vert
#version 330

in vec3 inPosition;

out vec3 outPosition;

vec3 velocities[3] = vec3[3](vec3(1, 0, 0), vec3(0, 1, 0), vec3(0, 0, 1));

void main()
{
    vec3 pos = inPosition + velocities[gl_VertexID];
    if (pos.x > 1 || pos.y > 1 || pos.z > 1)
    {
        pos = vec3(0, 0, 0);
    }

    outPosition = pos;

    gl_Position = vec4(pos, 1);
}
```

上一小节中,将第二个 VertexBuffer 绑定到 Transform Feedback Object 上,实现了更新第二个 VertexBuffer 的功能。如果将两个 VertexBuffer 分别绑定到两个 Transform Feedback Object 上,就可以实现轮流更新对方的功能。注意,这两个 VertexBuffer 也都被绑定到了 ShaderProgram 的输入变量上。

一个 VertexBuffer 可以视为下一帧的 VertexBuffer,即两个 VertexBuffer 描述的是同一属性在相邻帧的状态。因此,模型中需要用 2 个 VertexBuffer 描述三角形的位置,代码如下:

```csharp
partial class DemoModel : IBufferSource
{
    public const string strPosition = "position";
    public const string strPosition2 = "position2";
    private VertexBuffer positionBuffer;
    private VertexBuffer positionBuffer2;

    private IDrawCommand drawCmd;

    private static readonly vec3[] positions = new vec3[3];

    #region IBufferSource 成员
```

```csharp
public IEnumerable<VertexBuffer> GetVertexAttribute(string bufferName)
{
    if (bufferName == strPosition)
    {
        if (this.positionBuffer == null)
        {
            this.positionBuffer = positions.GenVertexBuffer(
                VBOConfig.Vec3, BufferUsage.DynamicCopy);
        }

        yield return this.positionBuffer;
    }
    else if (bufferName == strPosition2)
    {
        if (this.positionBuffer2 == null)
        {
            this.positionBuffer2 = positions.GenVertexBuffer(
                VBOConfig.Vec3, BufferUsage.DynamicCopy);
        }

        yield return this.positionBuffer2;
    }
    else
    {
        throw new ArgumentException("bufferName");
    }
}

public IEnumerable<IDrawCommand> GetDrawCommand()
{
    if (this.drawCmd == null)
    {
        this.drawCmd = new DrawArraysCmd(DrawMode.Triangles, positions.Length);
    }

    yield return this.drawCmd;
}

#endregion
}
```

为了轮流更新对方,需要 2 个 RenderMethod,分别以一个 VertexBuffer 作为输入数据,另一个保存输出数据。为了分别渲染 2 个 VertexBuffer,也需要 2 个 RenderMethod,即一共需要 4 个 RenderMethod,代码如下:

```csharp
partial class DemoNode : ModernNode, IRenderable
{
    public static DemoNode Create()
    {
        RenderMethodBuilder updateBuilder, updateBuilder2;
        {
            var vs = new VertexShader(updateVert);
            var feedbackVaryings = new string[] {"outPosition", };
            var array = new ShaderArray(feedbackVaryings, GL_SEPARATE_ATTRIBS, vs);
            var map = new AttributeMap();
            map.Add("inPosition", DemoModel.strPosition);
            var map2 = new AttributeMap();
            map2.Add("inPosition", DemoModel.strPosition2);
            updateBuilder = new RenderMethodBuilder(array, map);
            updateBuilder2 = new RenderMethodBuilder(array, map2);
        }

        RenderMethodBuilder renderBuilder, renderBuilder2;
        {
            var vs = new VertexShader(renderVert);
            var fs = new FragmentShader(renderFrag);
            var array = new ShaderArray(vs, fs);
            var map = new AttributeMap();
            map.Add("inPosition", DemoModel.strPosition);
            var map2 = new AttributeMap();
            map2.Add("inPosition", DemoModel.strPosition2);
            renderBuilder = new RenderMethodBuilder(array, map);
            renderBuilder2 = new RenderMethodBuilder(array, map2);
        }

        var model = new DemoModel();
        var node = new DemoNode(model,// IBufferSource
            updateBuilder, updateBuilder2, renderBuilder, renderBuilder2); // builders
        node.Initialize();

        return node;
    }

    private DemoNode(IBufferSource model, params RenderMethodBuilder[] builders)
        : base(model, builders)
    {
    }
}
```

Transform Feedback Object

为了得到输出数据，需要 2 个 Transform Feedback Object，分别绑定到 2 个 VertexBuffer 上，代码如下：

```csharp
class DemoNode : ModernNode, IRenderable
{
    private TransformFeedbackObject[] transformFeedbackObjects = new TransformFeedbackObject[2];

    protected override void DoInitialize()
    {
        base.DoInitialize();

        for (int i = 0; i < 2; i++)
        {
            var tfo = new TransformFeedbackObject();
            RenderMethod method = this.RenderUnit.Methods[i];
            // make sure there is only one vao in this case
            var vao = method.VertexArrayObjects[0];
            VertexShaderAttribute[] attributes = vao.VertexAttributes;
            for (uint bindingPoint = 0; bindingPoint < attributes.Length; bindingPoint++)
            {
                tfo.BindBuffer(bindingPoint, attributes[bindingPoint].Buffer);
            }
            this.transformFeedbackObjects[i] = tfo;
        }
    }
}
```

渲染时，需要用一个 updateMethod 来更新数据，然后用一个 renderMethod 来渲染数据。下次渲染时，用另一套 updateMethod 和 renderMethod 来交替，从而实现了不断地更新数据的目的，代码如下：

```csharp
classDemoNode : ModernNode, IRenderable
{
    #region IRenderable 成员

    Int currentIndex = 0;
    public void RenderBeforeChildren(RenderEventArgs arg)
    {
        TransformFeedbackObject tfo = transformFeedbackObjects[(currentIndex + 1) % 2];
```

```
        // update
        {
            glEnable(GL_RASTERIZER_DISCARD);

            RenderMethod method = this.RenderUnit.Methods[currentIndex];
            ShaderProgram program = method.Program;
            // set the uniform variables
            program.SetUniform(..);
            method.Render(tfo); // update buffers that bound to tfo's binding point

            glDisable(GL_RASTERIZER_DISCARD);
        }
        // render
        {
            RenderMethod method = this.RenderUnit.Methods[(currentIndex + 1) % 2 + 2];
            ShaderProgram program = method.Program;
            program.SetUniform(..); // setup uniforms
            //unit.Render() // this method requires specified vertexes' count
            tfo.Draw(method);
                        // render updated buffers without specifying vertexes' count
        }
        // exchange
        {
            currentIndex = (currentIndex + 1) % 2;
        }
    }

    #endregion
}
```

本示例的效果如图 14-1 所示。

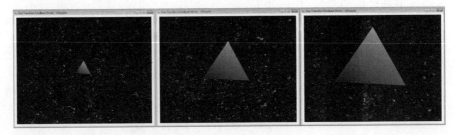

图 14-1　双 TFO 更新顶点缓存

14.4 粒子系统

在 2 个 Transform Feedback Object 轮流更新数据的基础上,将每一个顶点视为一个粒子,设计一个更新顶点位置(颜色等其他属性)的算法,就可以实现简单的粒子系统。读者可在本书网络资料的 c14d03_ParticleSystem 项目中找到本小节的全部代码。一个简单的更新粒子位置的 Shader 代码如下:

```
// vertex shader
#version 330

in vec3 inPosition; // position at this moment
in vec4 inVelocity; // velocity(xyz) and life(w) at this moment

uniform vec3 gravity = vec3(0, -9.81, 0);
uniform float deltaTime; // in seconds

out vec3 outPosition; // position at next moment
out vec4 outVelocity; // velocity(xyz) and life(w) at next moment

// return random value
float rand(float seed){
}

void main() {
    outPosition = inPosition + deltaTime * inVelocity.xyz;
    outVelocity.xyz = inVelocity.xyz + deltaTime * gravity;
    outVelocity.w = inVelocity.w - deltaTime;

    if (outVelocity.w < 0) // if life ends
    {
        outVelocity.w = 6; // reset life
        outVelocity.xyz = vec3(rand(..), rand(..), rand(..));
        outPosition = vec3(0, 5, 0) - vec3(rand(..), rand(..), rand(..)));
    }

    if (outPosition.y < -3) { // if particle touches ground
        outVelocity.xyz = reflect(outVelocity.xyz, vec3(0, 1, 0)) * 0.8;
        outPosition.y = -3;
    }
}
```

此 Shader 的作用是,计算在重力作用下的粒子在下一时刻的位置和速度。如果粒子碰撞到了地面,会被地面反弹起来。

为了将单个顶点渲染为漂亮的图案,在本例中使用了 Billboard 技术。首先启用

Geometry Shader,将1个顶点扩展为4个顶点构成的正方形,然后在 Fragment Shader 中根据 Fragment 相对正方形中心的距离涂上渐变色,代码如下:

```glsl
// vertex shader
#version 330

in vec3 inPosition;
in vec4 inVelocity;

out float life;

void main() {
    gl_Position = vec4(inPosition, 1.0);
    life = inVelocity.w;
}


// --------------------------------------------------
// geometry shader
#version 330

layout (points) in;
layout (triangle_strip, max_vertices = 4) out;

uniform mat4 projectionMat;
uniform mat4 viewMat;

in float life[];

out vec2 texCoord;
out float passLife;

void main() {
    passLife = life[0];

    vec4 pos = viewMat * gl_in[0].gl_Position;
    texCoord = vec2(-1, -1);
    gl_Position = projectionMat * (pos + 0.2 * vec4(texCoord, 0, 0));
```

```glsl
    EmitVertex();
    texCoord = vec2(1, -1);
    gl_Position = projectionMat * (pos + 0.2 * vec4(texCoord, 0, 0));
    EmitVertex();
    texCoord = vec2(-1,1);
    gl_Position = projectionMat * (pos + 0.2 * vec4(texCoord, 0, 0));
    EmitVertex();
    texCoord = vec2(1, 1);
    gl_Position = projectionMat * (pos + 0.2 * vec4(texCoord, 0, 0));
    EmitVertex();

    EndPrimitive();
}

// ----------------------------------------------------------
// fragment shader
#version 330

in vec2 texCoord;
in float passLife;

out vec4 outColor;

uniform vec4 color1 = vec4(0.3, 0.7, 0.1, 0.4);
uniform vec4 color2 = vec4(0.8, 0.2, 0.9, 0.1);

void main() {
    float distance = dot(texCoord, texCoord);
    if (distance > 0.5) discard;
    vec4 color = color1 * distance + color2 * (1.0 - distance);
    color.a *= (passLife / 6);
    outColor = color;
}
```

最后渲染 6000 个粒子的效果如图 14-2 所示。

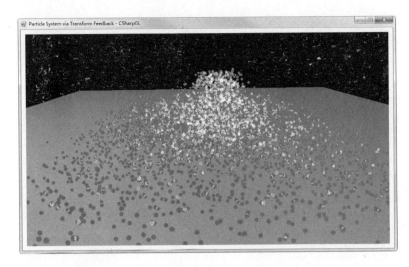

图 14 – 2 用 Transform Feedback 实现的粒子系统

14.5 总　结

本章介绍了 Transform Feedback Object 和一个实现粒子系统的方法。
① Transform Feedback 的作用？
在 GPU 端并行地更新顶点缓存的数据。
② Transform Feedback 实现粒子系统的思路？
2 个 Transform Feedback Object 轮流更新顶点缓存。

14.6 问　题

带着问题实践是学习 OpenGL 的最快方式。这里给读者提出问题，作为抛砖引玉之用。
请尝试修改粒子系统的 Shader，获得更绚丽的粒子效果。

第 15 章

Compute Shader

学完本章之后,读者将能:
- 理解使用 Compute Shader 的 Pipeline;
- 认识和使用 Compute Shader;
- 使用 Compute Shader 功能实现简单的粒子系统。

15.1 Compute Shader 简介

GPU 的并行计算能力十分强大,除了用于渲染三维场景,还具有解决其他问题的潜力。Compute Shader 就是借助 GPU 的并行计算能力解决通用问题的一种手段。Compute Shader 不与其他 Shader 配合使用,不参与渲染三维场景的 Pipeline。它由自己的调用命令来启动,即

```
// execute Compute Shader
void glDispatchCompute(uint num_groups_x, uint num_groups_y, uint num_groups_z);
```

这类似 Pipeline 中的 glDrawArrays(..) 和 glDrawElements(..) 函数。Compute Shader 将全部并行计算的工作划分为若干个小组,称为本地工作组(local workgroup)。在 glDispatchCompute(..) 函数中的参数分别指定了在 X、Y 和 Z 维度上的本地工作组的数量,每个参数都必须大于 0 且不超过 GL_MAX_COMPUTE_WORK_GROUP_SIZE(这是一个与设备相关的常量)。

每个本地工作组所需要做的工作由若干次 Compute Shader 的执行来完成,每一次执行被称为一个工作项(work item)。一个简单的 Compute Shader 代码如下:

```
//compute shader
#version 430 core

// size of a local work group
layout (local_size_x = 32, local_size_y = 16, local_size_z = 1) in;
layout (binding = 0, rgba32f) uniform image2D outImage;

// execute once for every work item
```

```
void main()
{
    ivec2 pos = ivec2(gl_GlobalInvocationID.xy);
    vec4color = vec4(vec3(gl_LocalInvocationID) / vec3(gl_WorkGroupSize), 1.0);
    imageStore(outImage, pos, color); // write to texture object.
}
```

此代码的作用是在一个纹理对象上的指定位置写入指定的颜色。其中的布局限定符 local_size_x、local_size_y 和 local_size_z 指定了本地工作组在 X、Y 和 Z 维度上的工作项的数量,其默认值均为 1。Compute Shader 会在全局工作组中的每个本地工作组的每个工作项上调用一次。

Compute Shader 中的内置变量记录了其在本地工作组或全局工作组中的位置(或者称为调用顺序)。Compute Shader 中的内置变量声明如下:

```
const uvec3 gl_WorkGroupSize;
```

上面这个常量的数值就是在布局限定符 local_size_x、local_size_y 和 local_size_z 中指定的数值。创建这个常量有两个目的。首先,它使得本地工作组的大小可以在 Shader 中被多次访问而无需依赖预处理。其次,它使得以多维形式表示的本地工作组大小可以直接按向量处理,而无须显式地构造。

```
in uvec3 gl_NumWorkGroups;
```

上面这个向量记录的是传递给 glDispatchCompute(..) 的参数(num_groups_x、num_groups_y 和 num_groups_z)。在 Shader 中可以直接用此变量获取全局工作组的大小。

```
in uvec3 gl_LocalInvocationID;
```

上面这个向量表示当前工作项在本地工作组中的位置,其范围从 uvec3(0) 到(gl_WorkGroupSize - uvec3(1))。

```
in uvec3 gl_WorkGroupID;
```

上面这个向量表示当前本地工作组在全局工作组中的位置,其范围从 uvec3(0) 到(gl_NumWorkGroups - uvec3(1))。

```
in uvec3 gl_GlobalInvocationID;
```

上面这个向量表示当前工作项在全局工作组中的位置,其值为(gl_WorkGroupID * gl_WorkGroupSize + gl_LocalInvocationID)。

```
in uint gl_LocalInvocationIndex;
```

上面这个数值是将 gl_LocalInvocationID 扁平化的结果,其值为[gl_LocalInvocationID.z * gl_WorkGroupSize.y * glWorkGroupSize.x + gl_LocalInvocation-ID.y * gl_WorkGroupSize.x + gl_LocalInvocationID.x]。由此可见,Compute Shader 的每个工作项的次顺是按如下伪代码安排的:

```
for (uint z = 0; z < gl_WorkGroupSize.z; z++)
{
    for (uint y = 0; y < gl_WorkGroupSize.y; y++)
    {
        for (uint x = 0; x < gl_WorkGroupSize.x; x++)
        {
            var workItem = new ComputeShader();
            workItem.gl_LocalInvocationID = new uvec3(x, y, z);
            workItem.gl_LocalInvocationIndex =
                z * gl_WorkGroupSize.y * glWorkGroupSize.x
                + y * gl_WorkGroupSize.x
                + x;
            workItem.Compute(); // execute compute shader.
        }
    }
}
```

之所以既在 Shader 中用布局限定符 local_size_x、local_size_y 和 local_size_z 指定工作项的数量,又在 glDispatchCompute(..) 的参数中指定有多少个这样的工作组,是因为在一个本地工作组内部存在更快速的读写方式。这样区分本地工作组和全局工作组是有意义的。

15.2 图像处理

本小节介绍几个用 Compute Shader 实现图像处理的示例。读者可在本书网络资料的 c15d00_InverseColor 项目中找到此示例的全部代码。在一个图像上,用 vec4(1,1,1,1)分别减去每个像素的颜色值,就可以得到反色的图像。实现图像反色滤镜的 Compute Shader 代码如下:

```
// compute shader
#version 430 core
//max width of bitmap is 512
layout (local_size_x = 512, local_size_y = 1, local_size_z = 1) in;
//input image
layout (binding = 0, rgba32f) uniform image2D inImage;
// result
layout (binding = 1, rgba32f) uniform image2D outImage;
```

```glsl
void main()
{
    ivec2 pos = ivec2(gl_GlobalInvocationID.xy);
    vec4 color = imageLoad(inImage, pos);
    vec4 inversedColor = vec4(1.0) - color;
    imageStore(outImage, pos, inversedColor);
}
```

在一个图像上,用一个像素四周的像素值相加取平均值,就可以得到一个简单的模糊滤镜。实现图像模糊滤镜的 Compute Shader 代码如下:

```glsl
// compute shader
#version 430 core
// max width of bitmap is 512
layout (local_size_x = 512, local_size_y = 1, local_size_z = 1) in;
//input image
layout (binding = 0, rgba32f) uniform image2D inImage;
// result
layout (binding = 1, rgba32f) uniform image2D outImage;

uniform int strength = 3;

void main()
{
    ivec2 pos = ivec2(gl_GlobalInvocationID.xy);
    vec4 color = vec4(0);
    for (int deltaX = -strength; deltaX <= strength; deltaX++)
    {
        for (int deltaY = -strength; deltaY <= strength; deltaY++)
        {
            int x = min(pos.x + deltaX, 511); x = max(x, 0);
            int y = min(pos.y + deltaY, 511); y = max(y, 0);
            color += imageLoad(inImage, ivec2(x, y));
        }
    }
    color /= ((strength * 2 + 1) * (strength * 2 + 1));

    imageStore(outImage, pos, color);
}
```

在一个图像上,将一个像素周围的像素的颜色取平均值,并赋予这些像素,就可以得到一个简单的马赛克滤镜。实现图像马赛克滤镜的 Compute Shader 代码如下:

```glsl
// compute shader
#version 430 core
// max width of bitmap is 512
layout (local_size_x = 8, local_size_y = 8) in;
//input image
layout (binding = 0, rgba32f) uniform image2D inImage;
// result
layout (binding = 1, rgba32f) uniform image2D outImage;

shared vec4 colors[64];

voidmain()
{
    ivec2 pos = ivec2(gl_GlobalInvocationID.xy);
    colors[gl_LocalInvocationIndex] = imageLoad(inImage, pos);
    barrier();

    vec4 result = vec4(0);
    for (int i = 0; i < 16; i++)
    {
        result += (colors[i] / 16);
    }

    imageStore(outImage, pos, result);
}
```

为避免多余的内存访问，本例用一个 shared 数组来存储输入图像的一小块，这个变量对同一个本地工作组内的所有工作项都可见。每个工作项都读取输入图像的目标像素，然后存储到 shared 数组。当本地工作组内的所有工作项都读取输入图像后，这个 shared 数组就含有输入图像当前这一小块的所有像素值。注意，barrier()函数使得本地工作组内的所有工作项都在此等待，直到它们都走到这一步为止。之后每个工作项都可以直接从此 shared 数组中读取像素值，这个读取速度是非常快的。这里就突显出全局工作组和本地工作组同时存在的价值：合理地配置本地工作组可以加速程序的执行。

通常读/写 shared 变量的速度远远高于读写图像或 Shader Storage Buffer，因为 Compute Shader 会将 shared 变量作为局部变量处理。因此，可以将内存中的数据复制到 shared 变量中，处理完后再写回到内存中。

Shared 变量位于高性能的资源环境中，这样的资源是有限的。需要查询一个 Compute Shader 支持的 shared 变量的最大数量时，可以以 GL_MAX_COMPUTE_SHARED_MEMORY_SIZE 为参数调用 glGetIntegerv(..)。

如果没有明确地指定本地工作组内工作项的执行顺序,也没有指定全局工作组中所有本地工作组的执行顺序,那么各个工作项的执行顺序是随机的。在需要读/写 shared 变量、图像、缓存或者其他共享的资源时,需要利用同步命令对某些操作进行同步控制。

一种同步命令是运行屏障(execution barrier),可以用 barrier() 函数触发。当某个工作项执行到 barrier() 函数时,它会暂停执行,直到它所在的本地工作组的所有工作项都执行到 barrier() 函数为止。基于此,必须在统一的控制流中调用 barrier(),以免工作项在暂停后无法重新启动。另一种同步命令是内存屏障(memory barrier),可以用 memoryBarrier() 触发。在一个工作项写入内存某处,然后使用 memoryBarrier(),在此之后读取此处的数据,必然是正常写入的数据,这就避免了读取脏数据。

memoryBarrier() 系列中有其他细化的函数版本。memoryBarrierAtomicCount() 函数会等待原子计数器更新完毕,然后再继续执行。memoryBarrierBuffer() 和 memoryBarrierImage() 函数会等待缓存和图像变量的写入操作完毕,memoryBarrierShared() 函数会等待标记为 shared 变量更新完毕。上述内存屏障函数都作用于全局工作组,而 groupMemoryBarrier() 只作用于单一的本地工作组。

用 Compute Shader 实现的这几种图像处理算法的效果如图 15-1 所示。

图 15-1 (左上)原始图像 (右上)模糊 (左下)反色 (右下)马赛克

15.3 粒子系统

本小节介绍一个用 Compute Shader 实现简单的粒子系统的示例，用 Compute Shader 来更新 1 百万个粒子的位置和速度。本示例的最终效果与 14.4 节的粒子系统相同，读者可在两个示例的相互对比中理解得更深入。读者可在本书网络资料的 c15d00_ParticleSystem 项目中找到本示例的完整代码。

首先分配 2 个大的缓存对象，一个存放粒子的位置，一个存放粒子的速度。系统开始运行后，Compute Shader 的每个工作项都只处理一个粒子。Compute Shader 从缓存中读取当前的速度和位置，计算出新的速度和位置，然后写入缓存中，周而复始，不断循环。粒子有一定的寿命，每次更新时都会减少；少到 0 时就会重置，且重置其位置，这样模拟过程就能持续地进行下去。用于更新粒子位置和速度的 Compute Shader 代码如下：

```glsl
// compute shader
#version 430 core

// 128 particles for each part
layout (local_size_x = 128) in;
//position(xyz) and life(w) of particles
layout (binding = 0, rgba32f) uniform imageBuffer positionBuffer;
//velocity of particles
layout (binding = 1, rgba32f) uniform imageBuffer velocityBuffer;

uniform vec3 gravity = vec3(0, -9.81, 0);
uniform float deltaTime; // in seconds
uniform float dieSpeed = 0.02;

// return random value
float rand(float seed){
}

void main()
{
    // read position and velocity of current particle.
    int id = int(gl_GlobalInvocationID.x);
    vec4 pos = imageLoad(positionBuffer, id);
    vec4 vel = imageLoad(velocityBuffer, id);
```

```glsl
    pos.xyz += deltaTime * vel.xyz;
    pos.w -= deltaTime;
    vel.xyz += gravity * deltaTime;

    if (pos.w < 0)
    {
        pos.w = 6; // reset life.
        vel.xyz = vec3(rand(..), rand(..), rand(..));
        pos.xyz = vec3(0, 5, 0) - vec3(rand(..), rand(..), rand(..));
    }

    if (pos.y < -3) { // if particle touches ground
        vel.xyz = reflect(vel.xyz, vec3(0, 1, 0));
        pos.y = -3;
    }

    // write position and velocity to buffer
    imageStore(positionBuffer, id, pos);
    imageStore(velocityBuffer, id, vel);
}
```

渲染粒子到画布的 Shader 与 14.4 节中的相同,不再重复。

注意,Vertex Shader 中使用的粒子位置属性,来自模型提供的 VertexBuffer 对象 positionBuffer。借助 OpenGL 函数 void glTexBuffer(uint target, uint internalFormat, uint buffer); 可以创建一个 Texture 对象 texPosition,并使其 Image 数据关联到 positionBuffer。这样,Compute Shader 使用的描述位置的缓存与 Vertex Shader 使用的位置属性缓存,指向了同一块内存区域,所以才能够直接在画布上渲染出粒子的运行状况。

15.4 总　结

本章介绍了 Compute Shader 及其两种用法:图像滤镜和粒子系统。
① Compute Shader 用于解决什么问题?
任何可并行化的计算问题。
② 如何确保 Compute Shader 不读写脏数据?
使用运行屏障或内存屏障。
③ 如何加快 Compute Shader 的执行效率?
使用 shared 变量,合理设置本地工作组的大小。

15.5 问　题

带着问题实践是学习 OpenGL 最快的方式。这里给读者提出几个问题,作为抛砖引玉之用。

1. 请运行本书网络资料的 c15d00_InverseColor 项目,试用不同的图片观察各个滤镜的效果。尝试调整算法以产生更好的效果。

2. 请在本书之前的章节查找使用了 glTexBuffer(..) 函数的情形,总结其功能用法。

附 录

附录 A　Github 入门

学完本章之后,读者将能:
- 使用 Github 同步保存代码;
- 用 Github 找回需要的代码版本;
- 用 Github 的 Branch 和 Pull Request 更新代码。

A.1　Github 简介

本书所说的 Github,是指一个网站(https://github.com)。Github 可以保存项目的所有代码、资源和文档等文件,能够记录每次更新的内容,支持多人共同维护一个项目。从最基本的角度,可以将其视为一个能够自动记录所有代码版本的网络硬盘。本章以使用 CSharpGL 为目的,只介绍 Github 的一小部分功能和用法,以及简单的操作方法。更多高级知识请参考其他资料。

如果读者下载速度偏慢,或者不想注册 Github 账号,可以到 CSharpGL 托管页面直接下载独立的代码压缩包,如图 A-1 所示。

图 A-1　下载独立的代码压缩包

独立的代码压缩包不能使用Github的托管功能,因此强烈建议读者按照本节介绍的步骤用Github Desktop下载和使用CSharpGL。

A.2 下载和安装

Github支持命令行和可视化两种操作方式。这里以可视化方式为例。首先,到https://desktop.github.com下载最新版的Github可视化工具软件安装包GitHubDesktopSetup.exe文件(或者直接在本书网络资料中查找此安装包),然后双击安装。安装过程十分简单,只有如图A-2所示的一个等候界面。

图 A-2 安装 Github Desktop

安装完毕后,Github Desktop会自动打开,界面如图A-3所示。

可以看到,在Github Desktop里可以新建代码仓库(repository),可以将本地的代码仓库附着到Github上,还可以将Github上的代码仓库复制到本地。

A.3 复制代码仓库

为了使用CSharpGL,这里将Github网站上的CSharpGL代码仓库复制到本地。如图A-3所示,单击右侧的Clone a repository按钮,由于还没有告知Github Desktop关于Github账号的信息,Github Desktop会要求用户使用Github账号登陆,如图A-4所示。

因此,必须先登陆Github网站注册一个账号,如图A-5所示。

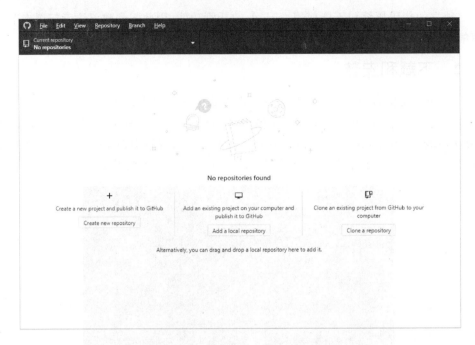

图 A-3　首次打开 Github Desktop

图 A-4　先登陆后才能复制代码仓库

图 A-5　注册一个 Github 账号

完成注册后,读者可以在 Github 上搜索"CSharpGL",或者直接输入网址 https://github.com/bitzhuwei/CSharpGL,找到 CSharpGL 的托管页面。在此页面右上角,单击 Fork 按钮,即可将此代码仓库复制到读者自己的 Github 账号上,如图 A-6 所示。

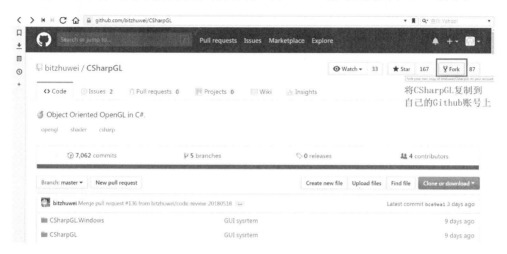

图 A-6　将别人的代码仓库复制到自己的账号

注意,如果想用 Github Desktop 将其他人的代码仓库复制到本地,必须先用 Fork 功能将其他人的代码仓库复制到自己的 Github 账号上。Github Desktop 只能将自己账号上的代码仓库复制到本地并进行管理。

现在,读者的 Github 账号上已经有了一份 CSharpGL,继续之前在 Github Desktop 上的操作,登陆 Github,如图 A-7 所示。

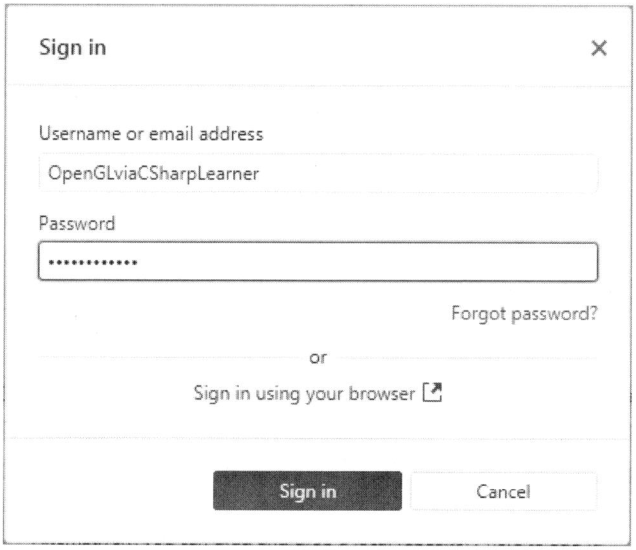

图 A-7　在 Github Desktop 上登陆 Github 账号

登陆成功后，再次单击 Clone a repository 按钮，Github Desktop 会自动获取读者 Github 账号上现有的所有代码仓库。选中 CSharpGL，设置合适的本地存储路径，单击 Clone 按钮，如图 A-8 所示。

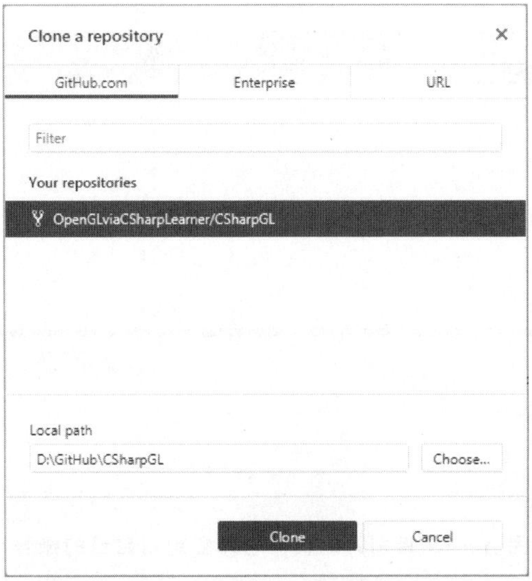

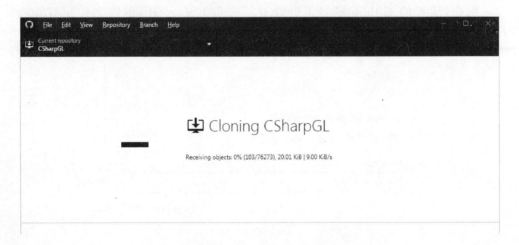

图 A-8　选则 CSharpGL 并复制到本地

Github Desktop 会开始下载 CSharpGL，如图 A-9 所示。

图 A-9　Github Desktop 正在下载 CSharpGL

下载完成后，可以在 Github Desktop 右侧界面的链接中直接打开本地代码的文件夹，如图 A-10 所示。

此时，可以像使用任何普通的代码一样使用本地的代码仓库。

附 录

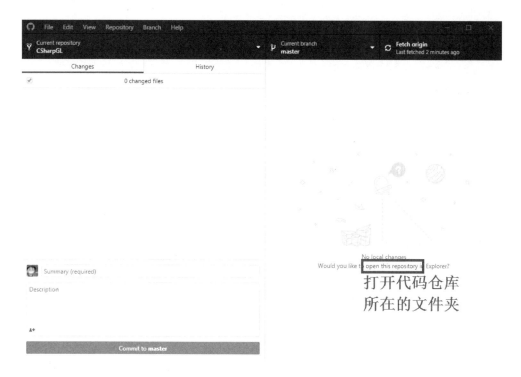

图 A-10 刚刚下载到本地的代码仓库

A.4 上传代码

使用 Github 的好处之一是,可以随时将本地的最新代码上传到 Github 保存,避免因本地代码意外丢失而造成损失。本节举例说明如何利用 Github Desktop 上传本地代码到 Github。

打开刚刚下载到本地的代码仓库,找到 c01d00_Cube\Form1.cs 文件,在其中一处新增代码如下:

```
private void FormMain_Load(object sender, EventArgs e)
{
    this.cubeNode = CubeNode.Create();
    this.cubeNode.Scale = new vec3(3f, 0.2f, 2.5f); // insert a new line
}
```

新增代码的功能是修改节点的缩放比例。保存文件后,会在 Github Desktop 中看到被修改的内容,如图 A-11 所示。

现在,确认这是期望修改的内容。然后,在 Github Desktop 左下方的 Summary (required)文本框内填写对此次修改的总结。一般的,此处应填写为何做此修改,或

图 A‑11　Github Desktop 展示所有修改过的文件

者修改了哪些内容等有用信息。填写完成后，单击左下角的 Commit to master 按钮，Github Desktop 就会确认完成一次修改。从修改代码到现在为止，就称作完成了一次在本地的 Commit。

此时，读者对代码所做的 Commit 操作都仅保留在本地。读者可以单击 Github Desktop 上方的 Push origin 按钮，将对本地代码仓库的所有 Commit，都上传到 Github 网站，如图 A‑12 所示。

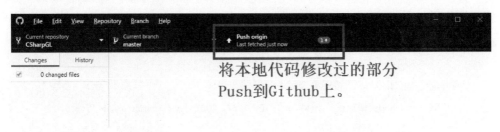

图 A‑12　将本地的 Commit 提交到 Github

稍候片刻，Github Desktop 就完成了同步本地代码仓库到 Github 网站的操作。

A.5 分 支

在需要对项目进行大的调整时,或者在需要为项目新增大规模的复杂功能时,往往不是提交一次或几次 Commit 就能够完成的。而且在此过程期间,常常会发现写错代码,需要从之前的某一状态开始重写。另外,在实际应用的项目中,开发者并不希望在新功能完成之前就影响到用户。为解决这些问题,可以使用 Github 提供的分支(branch)功能。

我在 Github 上创建 CSharpGL 代码仓库时,Github 自动为其添加了一个分支 master,即主分支。读者在用 Fork 功能将 CSharpGL 复制到自己的代码仓库时,也复制了分支 master。在上一节中,所有的 Commit 操作都是在 master 分支上进行的。

如果需要为项目新增某项大型功能,可以在开始编码前为此功能创建一个分支,然后在此分支上进行编码,提交 Commit。在这一过程中,master 分支上的内容是不变的。等新分支上的功能编写测试完毕后,可以创建一个 Pull Request,将新分支上修改的内容一次性提交到 master 分支上。可以简单地认为,新分支就是一个包含了很多 Commit 的包裹。本节通过一个简单的示例演示如何使用分支。

A.5.1 创建分支

首先,单击 Github Desktop 界面上方的 Current branch 按钮,在弹出的 Branches 栏目下的文本框中填写想要新建的分支名称,并单击 New branch 按钮,如图 A-13 所示。

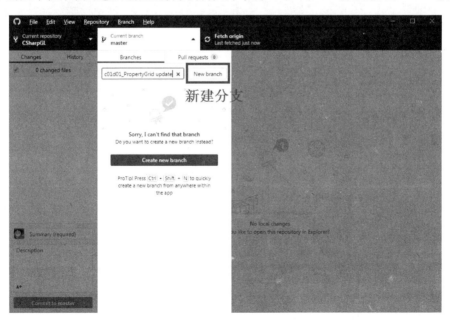

图 A-13 输入新分支名称

Github Desktop 会根据读者输入的分支名称做必要的修改,保证分支名称符合要求,并且提醒读者新分支是基于 master 分支的,如图 A-14 所示。

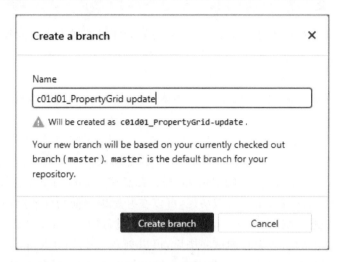

图 A-14 新建分支

单击 Create branch 按钮,Github Desktop 会创建新的分支,并使之成为当前使用的分支,如图 A-15 所示。

图 A-15 新分支创建完毕

A.5.2 上传代码

新分支创建完成后,就可以在新分支下修改代码,提交 Commit。这些操作与在 master 分支下的操作是相同的。示例中将 c01d01_PropertyGrid 项目中的代码进行了修改,如图 A-16 所示。然后,在此项目下新增文件 readme.md,用于描述此项目,如图 A-17 所示。

每完成一步修改,就可以提交一个 Commit 到本地代码仓库。所有的 Commit 都需要通过 Push origin 按钮提交到 Github 上。此时,Github 上的新分支的内容就与本地同步了。

A.5.3 创建 Pull Request

当新增的复杂功能完成编码测试并且可以发布后,可以简单地通过创建 Pull Request 的方式将新分支合并到 master 分支上,这就实现了发布新功能的目的。

图 A-16　在新分支下修改代码文件

图 A-17　在新分支下新增文件

单击 Current branch 按钮,在下方呈现出一个(create a pull request)链接。单击此链接,Github Desktop 会打开本地默认的浏览器并准备好创建 Pull Request,如图 A-18 所示。

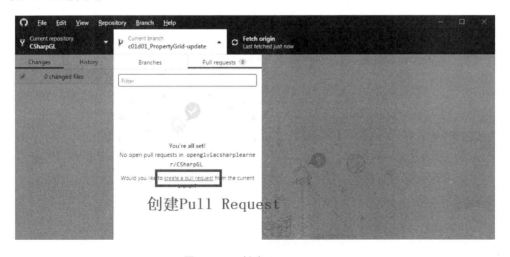

图 A-18　创建 Pull Request

浏览器页面呈现出创建一个 Pull Request 所需的各个选项,如图 A-19 所示。
注意,读者是在自己的项目下创建 Pull Request,要将选项中的 base:master 设置为自己项目下的 master。单击 Create pull request 按钮,即可完成创建 Pull Re-

图 A-19　填写相应内容并创建 Pull Request

quest。页面会跳转到合并页，如图 A-20 所示。

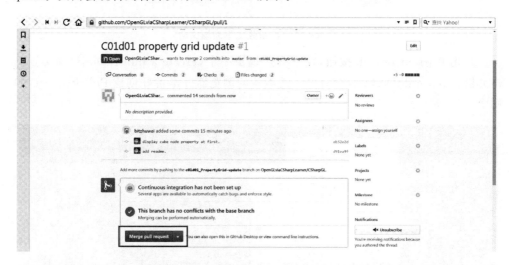

图 A-20　将新分支上的 Pull Request 合并到 master 分支

如果读者在新分支上修改的代码与 Github 的 master 分支上的代码没有冲突，Github 就会呈现 Merge pull request 按钮。单击此按钮，即可将新分支上的所有 Commit 都复制到 master 分支上。注意，这里的代码没有冲突，并不是指代码是可成功编译的，而是指对代码的所有 Commit 中的修改，是不冲突的。即使代码不能成功编译，仍旧可能是不冲突的。此时，新分支的任务已经完成，应该删除此分支。Github 页面也呈现出删除分支的按钮 Delete branch。单击此按钮，新分支将被删除，如图 A-21 所示。

此时 Github 上的新分支被删除，所有新功能的代码都存在于 master 分支上。

附 录

图 A-21 删除新分支

而本地代码仓库还处于新分支的状态,因此必须将其调回 master 分支。

A.5.4 同步代码至本地

现在 Github 网站上的代码仓库版本比本地仓库的版本更新,所以需要将 Github 网站上的代码同步至本地。在 Github Desktop 上单击 Current branch 按钮,在下方呈现的分支中选中 master,如图 A-22 所示。

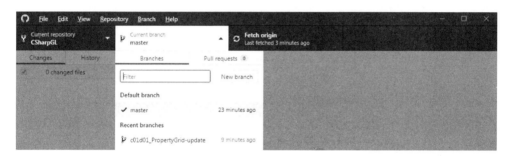

图 A-22 选中 master 分支

之后 Github Desktop 会自动检查本地代码仓库与 Github 网站上的版本差异。单击 Pull origin 按钮,即可将最新的远程代码复制到本地代码仓库,如图 A-23 所示。

图 A-23 将最新的远程代码复制到本地

注意,Pull origin 按钮右侧的数值"3+",表示在远程的 Github 上的 master 分支的代码仓库比本地代码仓库多3个Commit,即远程的代码版本更新,因此同步操作会用远程的代码覆盖本地代码。

现在,本地代码再次与Github网站上的代码保持一致。如果读者打开Github Desktop左侧的History栏,会看到Ahead子栏目下出现了新增的各个Commit。Ahead意味着这些Commit是读者创建的,并且与笔者的CSharpGL代码仓库相比,新增了这些Commit,如图A-24所示。

图A-24 Fork的CSharpGL项目下的Ahead于笔者的Commit

如果读者愿意,可以通过Pull Request的方式,将这些Ahead的Commit提交给我;如果我愿意,可以将读者的这些Commit合并到我的代码仓库。提交这一Pull Request的流程与上一节的流程相似,只是要将提交目标base:master修改为我的base:master。这一读者提交给我,然后我合并到master的过程,实际上就是Github实现团队合作开发的基本流程。在这一流程中,可以接受、修改和关闭Pull Request。Github在团队成员的交流中,保证了代码的一致性。

A.6 提问和讨论

Github不仅是一个在网络上保存和更新代码的网站,还是一个供全球程序员交流的场所。如果读者对CSharpGL这个项目有任何问题或建议,就可以在CSharpGL的页面上创建一个issue来讨论,如图A-25所示。

这类似于在论坛发帖,读者可以发表任何与项目相关的内容。读者与其他人(包括我)都可以相互回帖。通过与世界各地的程序员相互交流经验和看法,能够极大地提升自己的见解。

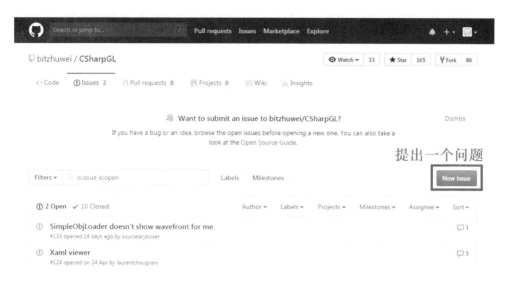

图 A-25　在 CSharpGL 页面提问

A.7　readme.md 文件

本章的示例中添加的 readme.md 是 Mark Down 格式的文件，可以简单地理解为一种简化的 HTML 文件，它支持显示图片、链接和代码等常见格式。

在 Github 网站上浏览代码时，Github 会自动展示当前代码文件夹下的 readme.md 文件的内容（如果存在）。因此 readme.md 文件是一个便于他人和自己理解当前文件夹下的代码的工具。所有比较复杂的代码文件夹下都应有这样一个 readme.md 文件。

附录 B　C♯和面向对象入门

学完本章之后，读者将能：
- 找到一种理解面向对象程序的方法；
- 用面向对象的思想理解 CSharpGL 和 OpenGL。

理解本附录的内容需要读者掌握 C 语言的结构体、指针、函数、函数指针和堆栈的内容。读者可以在任何一本关于 C 语言的入门书籍资料中获得相关信息。

理解和学会运用面向对象程序设计是困难的，因为面向对象是一种思想，是一种思维方式，是抽象的东西。仅仅通过比喻或类比的讲述方式无法揭示其精妙之处。只有经历过大量且丰富的代码的洗礼才能自发地找到面向对象的思维方式。如果说通过本章能够让读者一步到位地掌握面向对象编程，那一定是假话。本章将用 C 语言拆解面向对象的代码，从而让读者能够找到一条将面向对象代码降解到简单的面向过程代码的路径。如果读者能够一眼看穿面向对象代码,那么距离理解和学会运用它就不再遥远了。

面向对象语言的三大特性(封装、继承、多态)中,封装和继承的 C 版写法需要研究,而多态不在乎新的写法。所以本章就展示如何用 C 来模拟 C♯ 的封装、继承、虚方法、关键字 as 和 interface 的写法。

B.1 类是升级版的结构体

例如如下的 C♯ 代码,写了一个典型的类(class):

```
classVehicle
{
    public float speed = 12; // field

    public void PrintSpeed() // method
    {
        Console.WriteLine("PrintSpeed(), speed({0})", this.speed);
    }
}
```

其典型的使用方式代码如下:

```
// create aVehicle object
Vehicle obj = new Vehicle();
// use Vehicle object's field
Console.Write("speed({0})\n", obj.speed);
// useVehicle object's method
obj.PrintSpeed();
```

如果用 C 来实现类似的使用方式,应该如何写呢?

B.1.1 用 struct 代替 class

用 C 的 struct 代替 C♯ 的 class 来封装字段,代码如下:

```
typedef struct _Vehicle
{
    float speed; // field in C style
}Vehicle;
// method in C style
void PrintSpeed(Vehicle * pThis) // pThis means the keyword this in c♯
{
    printf("PrintSpeed(), speed(%f)\n", pThis->speed);
}
```

struct 没有"在类型里面写方法"的这种功能,因此要用变通的方法来实现。也就是说,用一个函数代替 class 里的方法,且这个函数的第一个参数是指向这个结构体的指针。

B.1.2 用 New[ClassName]代替 new

一个 class 类型至少有一个构造方法,用于 new 一个对象。所以还需要写一个 C 版的 new 方法,代码如下:

```
Vehicle* NewVehicle()
{
    // alloc memory
    Vehicle* obj = (Vehicle*)malloc(sizeof(Vehicle));
    // initialize fields
    obj->speed = 12;

    return obj;
}
```

C 版的 New 方法用 malloc 来申请内存空间,这样就把对象放到了堆上,与面向对象语言的处理方式相同。其使用方法代码如下:

```
//C#: Vehicle obj = new Vehicle()
Vehicle* obj = NewVehicle();
// C#: Console.Write("speed({0})\n", obj.speed)
printf("speed(%f)\n", obj->speed);
//C#: obj.PrintSpeed()
PrintSpeed(obj);
```

与 C#版的使用方法异曲同工。

B.2 继承和多态的本质是操纵函数指针

在上一节的 Vehicle 基础上,来讨论如下代码的实现:

```
class Car :Vehicle
{
    public float gas = 34;
    public void PrintGas()
    {
        Console.WriteLine("PrintGas(), gas({0})", this.gas);
    }
}
```

B.2.1 用组合代替继承

class 类型仍然用 struct 代替。Car 继承了 Vehicle 这一性质,如何实现呢？C 语言虽然没有"继承"的概念,但是 struct 可以组合(与 C♯ 的组合概念一样),这里用一个 Vehicle 的指针来表示继承的概念,代码如下:

```c
typedef struct _Car
{
    // base type
    Vehicle * pBase;
    // own fields
    int gas;
} Car;

Car * NewCar()
{
    // initialize base type
    Vehicle * pBase = NewVehicle();
    // alloc memory
    Car * obj = (Car *)malloc(sizeof(Car));
    obj->pBase = pBase;
    // initialize fields
    obj->gas = 34;

    return obj;
}

void PrintGas(Car * pThis)
{
    printf("PrintGas(), gas(%f)\n", pThis->gas);
}
```

在 C 版的构造函数 NewCar() 里,首先创建了一个"基类"Vehicle 的对象,然后赋值给"子类"Car 的对象的 pBase 指针。

这里也符合面向对象语言"先调用基类的构造函数,再调用子类的构造函数"的特点。具体地说,是:首先,子类的构造函数调用基类的构造函数;然后,完成对基类构造函数的调用;最后,完成对子类构造函数的调用。因此,一个子类对象实际上是由自身的 struct 和基类的 struct 共同组成的,如图 B-1 所示。

子类型 Car 的典型使用方式如下:

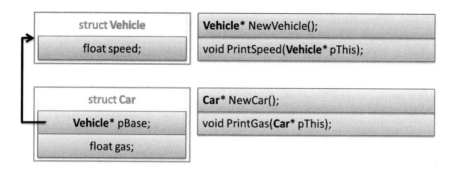

图 B-1 用组合表示继承的 Car 类型的对象

```
// create a Car object
Car obj = new Car();
// use Car object's field
Console.Write("gas({0})\n", obj.gas);
// use Car object's basetype's field
Console.Write("speed({0})\n", obj.speed);
// use Car object's method
obj.PrintGas();
// use Car object's basetype's method
obj.PrintSpeed();
```

C 版的子类型 Car 使用方式,仍然与此类似,代码如下:

```
//C#: Car obj = new Car();
Car * obj = NewCar();
//C#: Console.Write("gas({0})\n", obj.gas);
printf("gas( % f)\n", obj ->gas);
//C#: Console.Write("speed({0})\n", obj.speed);
printf("speed( % f)\n", obj ->pBase ->speed);
//C#: obj.PrintGas();
PrintGas(obj);
//C#: obj.PrintSpeed();
PrintSpeed(obj ->pBase);
```

这里可以看到,"子类"Car 能够调用自身的字段和方法,也能够借助 pBase 指针调用基类的字段和方法,完全达到了面向对象语言的要求。

B.2.2 用函数指针代替 virtual

继承还有一个特性,就是虚方法。C 语言如何做到呢? 答案就是使用函数指针。为展示清楚,这里重新定义上一节的两个类,代码如下:

```
class Vehicle
{
    public virtual void PrintSelf() // virtual method
    {
        Console.Write("a transport machine.\n");
    }
}
class Car :Vehicle
{
    public override void PrintSelf() // overrided method
    {
        Console.Write("a car.\n");
    }
}
```

其典型的使用方式代码如下：

```
// create aVehicle object
Vehicle vObj = new Vehicle();
vObj.PrintSelf();
// create a Car object
Car oObj = new Car();
oObj.PrintSelf();
// Car object assigned to Vehicle type
Vehicle vObj2 = new Car();
vObj2.PrintSelf();
```

其输出应该是：

```
a transport machine.
a car.
a car.
```

虚函数的特点，在于基类知道子类的override函数在哪儿。如果在创建子类的时候改变基类能够调用的函数，就可以做到这一点。所以很自然地就需要用到函数指针。按照上一节的方法，在New[ClassName]创建基类结构体的时候，让基类结构体的函数指针指向基类的virtual方法。等基类结构体创建完毕，继续创建子类结构体的时候，修改基类结构体的函数指针，使其指向子类的override方法，这就要求基类和virtual方法和子类的override方法的声明完全相同。其对应的C版代码如下：

```c
typedef struct _Vehicle
{
    void ( * PrintSelf )(_Vehicle * ); // function pointer for virtual method
}Vehicle;

void PrintSelf_Vehicle(Vehicle * pThis); // virtual method declaration

Vehicle * NewVehicle()
{
    // initialize base type
    // alloc memory.
    Vehicle * obj = (Vehicle * )malloc(sizeof(Vehicle));
    // initialize fields
    // initialize virtual methods
    obj ->PrintSelf = PrintSelf_Vehicle;

    return obj;
}

void PrintSelf_Vehicle(Vehicle * pThis) // virtual method implementation
{
    printf("transport machine.\n");
}

typedef struct _Car
{
    Vehicle * pBase;
} Car;

void PrintSelf_Car(Vehicle * pThis); // overrided method declaration

Car * NewCar()
{
    // initialize base type
    Vehicle * pBase = NewVehicle();
    // alloc memory
    Car * obj = (Car * )malloc(sizeof(Car));
    obj ->pBase = pBase;
    // initialize fields
    // initialize virtual methods
    pBase ->PrintSelf = PrintSelf_Car; // method pointer binding to overrided method
```

```
        returnobj;
}

void PrintSelf_Car(Vehicle * pThis) // overrided method implementation
{
        printf("a car.\n");
}
```

C版代码描述的虚函数的指针,在"基类"的构造函数中指向的是"基类"的函数,但在"子类"的构造函数中,会被修改为指向"子类"的函数,如图B-2所示。

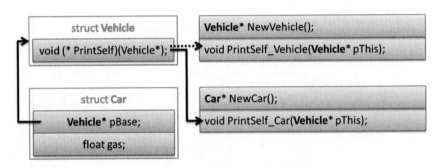

图B-2 用函数指针实现虚函数的Car类型的对象

C版的使用方法代码如下:

```
//C#: Vehicle vObj = new Vehicle()
Vehicle * vObj = NewVehicle();
// C#: vObj.PrintSelf()
vObj ->PrintSelf(vObj);
//C#: Car oObj = new Car()
Car * oObj = NewCar();
// C#: oObj.PrintSelf()
oObj ->pBase ->PrintSelf(oObj ->pBase);
//C#: Vehicle vObj2 = new Car()
Vehicle * vObj2 = NewCar() ->pBase;
// C#: vObj2.PrintSelf()
vObj2 ->PrintSelf(vObj2);
```

其使用方式异曲同工,而输出和C#版是一样的。

如果基类有多个virtual方法且其声明相同,就可以使用函数指针数组;有几种不同声明的virtual方法,基类就要有几个函数指针(或函数指针数组)指向他们。

B.2.3 用Convert2Type代替as

看如下使用关键字as的示例代码:

```
classVehicle
{
}
class Car :Vehicle
{
    public int gas = 34;
    public void PrintGas()
    {
        Console.Write("gas({0})\n", this.gas);
    }
}
```

其使用方式代码如下：

```
Vehicle baseObj = new Car();
Car derivedObj = baseObj as Car; // from base type to derived type
if (null != derivedObj)
{
    derivedObj.PrintGas();
}
```

上一节展示了在 C 中，如何用基类的指针指向子类的对象。而这里的 C# 的关键字 as 将基类的指针还原为子类的指针，这是如何实现的？关键字 as 可以用一个函数代替，函数名叫做 Convert2Type（或者任何读者喜欢的名字），as 的本质就是一个函数。问题是：已知一个对象的指针，已知要转换出来的目标类的类型，求出该对象指向目标类的指针；若不存在，则返回 NULL。

首先需要一些准备工作。类似数据库的主键，每个类型都要有一个唯一的标识符（用 int 即可），用于区分不同的类型。代表基类对象的 struct 要有一个指向子类对象的指针，由于子类可能有多种，显然只能用 void * 类型。

每一个类型都要记录基类指针、子类指针和唯一标识符等内容。另外，Convert2Type 函数只能有一个声明，不可能把所有类型的指针都传给它。为了给它尽可能多的数据，应该将类型的基类指针、子类指针、唯一标识符等信息封装为一个单独的 struct Metadata。具有继承关系的两个类型，其 Metadata。对象也相互指向。这样，Metadata 就包含了描述全部继承关系的数据，代码如下：

```
typedef struct _Metadata
{
    void * pThis; // point to struct of 'this' object
    int typeId; // identifier of 'this' object
    _Metadata * pBase; // point to metadata of base object
```

```c
    _Metadata * pDerived; // point to metadata of derived object
} Metadata;

Metadata * NewMetadata(
    void * pThis, // 'this' object
    int typeId, // type id
    Metadata * pBase) // info struct of base type
{
    Metadata * result = (Metadata *)malloc(sizeof(Metadata));

    result->pThis = pThis;
    result->typeId = typeId;
    result->pBase = pBase;
    result->pDerived = NULL;

    if (NULL != pBase)
    {
        pBase->pDerived = result; // base type points to this
    }

    return result;
}
```

为方便说明Convert2Type的实现原理,这里再定义C版的Vehicle和Car,代码如下:

```c
typedef struct _Vehicle
{
    // basic info
    Metadata * metaInfo;
    // fields

    // methods

}Vehicle;

// type id
static int VehicleTypeId = 4;

// method declarations
// the new method
Vehicle * NewVehicle()
```

```c
{
    // alloc memory
    Vehicle * obj = (Vehicle * )malloc(sizeof(Vehicle));

    // initialize basic info
    obj->metaInfo = NewMetadata(obj, VehicleTypeId, NULL);
    // initialize fields
    // initialize methods

    returnobj;
}

typedef struct _Car
{
    // basic info
    Metadata * metaInfo;

    // fields
    int gas;

    // methods

} Car;

// type id
static int CarTypeId = 5;

// method declarations
void PrintGas(Car * pThis);

// the new method
Car * NewCar()
{
    // initialize base type
    Vehicle * pBase = NewVehicle();

    // alloc memory
    Car * obj = (Car * )malloc(sizeof(Car));
    // initialize basic info
    obj->metaInfo = NewMetadata(obj, CarTypeId, pBase->metaInfo);

    // initialize fields
```

```
    obj->gas = 34;

    // initialize methods

    returnobj;
}

void PrintGas(Car * pThis)
{
    printf("gas( %d)\n", pThis->gas);
}
```

现在，Metadata 持有基类、自身、子类的信息，将其作为参数传入 Convert2Type 函数，是能够找到目标类型的，代码如下：

```
void * Convert2Type(Metadata * pThisMetaInfo, int targetTypeId)
{
    if (NULL == pThisMetaInfo || targetTypeId < 0) { return NULL; }

    void * result = NULL;

    Metadata * pCurrent = pThisMetaInfo->pDerived;// search in derived types
    while (pCurrent != NULL)
    {
        if (pCurrent->typeId == targetTypeId) // this is the type we are looking for
        {
            result = pCurrent->pThis;
            break;
        }

        pCurrent = pCurrent->pDerived;
    }

    pCurrent = pThisMetaInfo; // search in base types
    while (pCurrent != NULL)
    {
        if (pCurrent->typeId == targetTypeId) // this is the type we are looking for
        {
            result = pCurrent->pThis;
            break;
        }
```

```
            pCurrent = pCurrent ->pBase;
        }

        return result;
}
```

关键字 as 通过 Convert2Type 函数和一个 C 语言的强制类型转换实现。此时的 Car 类型的对象如图 B-3 所示。

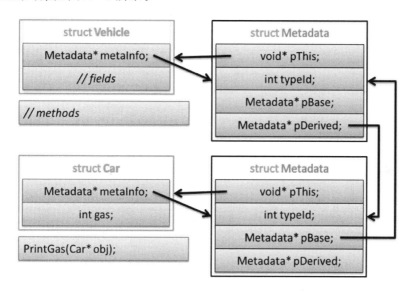

图 B-3 用 Metadata 维护内部关联的 Car 类型的对象

B.2.4 用链表代替 interface

C♯中的一个类可以实现多个 interface，这在 C 中如何表达？答案是：interface 也用 struct 代替。Interface 只不过是一个特殊的 class 类型，其内部字段都是函数指针。C♯的 class 类型实现一个 interface 类型，实际上仍然是继承了这个 interface 类型。这和继承基类类型没有区别。但一个 class 可以实现多个 interface，这就需要用链表来记录这些 interface，代码如下：

```
typedef struct _LinkNode
{
    _Metadata * pValue; // point to metadata of an interface
    _LinkNode * pNext; // point to next LinkNode
} LinkNode;

LinkNode * NewLinkNode(Metadata * interfaceMeta)
{
    LinkNode * result = (LinkNode * )malloc(sizeof(LinkNode));
```

```
    result->pValue = interfaceMeta;
    result->pNext = NULL;

    return result;
}
```

其中的 void * pValue;用于保存一个 class 类型实现的 interface 的 Metadata。为 Metadata 添加用于记录 interface 的链表指针,代码如下:

```
typedef struct _Metadata
{
    void * pThis; // point to struct of 'this' object
    int typeId; // identifier of 'this' object
    _Metadata * pBase;// point to metadata of base object
    _Metadata * pDerived;// point to metadata of derived object
    LinkNode * pInterfaceList;// list of interfaces this type supports
} Metadata;

Metadata * NewMetadata(
    void * pThis, // 'this' object
    int typeId, // type id
    Metadata * pBase, // info struct of base
    LinkNode * firstInterfaceNode) // list of interfaces this type supports
{
    Metadata * result = (Metadata *)malloc(sizeof(Metadata));

    result->pThis = pThis;
    result->typeId = typeId;
    result->pBase = pBase;
    result->pDerived = NULL;
    result->pInterfaceList = firstInterfaceNode;

    if (NULL != pBase)
    {
        pBase->pDerived = result; // base type points to this
    }

    LinkNode * node = firstInterfaceNode;
    while (NULL != node)
    {
        node->pValue->pDerived = result; // interface type points to this
        node = node->pNext;
    }

    return result;
}
```

相应地,需要修改 Convert2Type(..)函数,代码如下:

```
void * Convert2Type(Metadata * pThisMetaInfo, int targetTypeId)
{
    if (NULL == pThisMetaInfo || targetTypeId < 0) { return NULL; }

    Metadata * pCurrent = pThisMetaInfo->pDerived;// search in derived types
    while (pCurrent != NULL)
    {
        if (pCurrent->typeId == targetTypeId) // this is the type we are looking for
        {
            return pCurrent->pThis;
        }

        // search in interfaces of pCurrent
        LinkNode * interfaceNode = pCurrent->pInterfaceList ;
        while (interfaceNode != NULL)
        {
            Metadata * interfaceMeta = interfaceNode->pValue;
            if (interfaceMeta->typeId == targetTypeId)
            {
                return interfaceMeta->pThis;
            }
            interfaceNode = interfaceNode->pNext;
        }

        pCurrent = pCurrent->pDerived;
    }

    pCurrent = pThisMetaInfo; // search in this and base types
    while (pCurrent != NULL)
    {
        if (pCurrent->typeId == targetTypeId) // this is the type we are looking for
        {
            return pCurrent->pThis;
        }

        // search in interfaces of pCurrent.
        LinkNode * interfaceNode = pCurrent->pInterfaceList ;
        while (interfaceNode != NULL)
        {
            Metadata * interfaceMeta = interfaceNode->pValue;
```

```
            if (interfaceMeta->typeId == targetTypeId)
            {
                return interfaceMeta->pThis;
            }

            interfaceNode = interfaceNode->pNext;
        }

        pCurrent = pCurrent->pBase;
    }

    return result;
}
```

下面来验证上述方案。创建一个 interface 类型,让 Vehicle 实现它,代码如下:

```
interface IMovable
{
    void Move();
}
classVehicle : IMovable
{
    public void Move()
    {
        Console.Write("Vehicle.Move()\n");
    }
}
```

C#版的使用方式如下:

```
Vehicle obj = new Vehicle();
obj.Move();
IMovable interfaceObj = obj as IMovable;
interfaceObj.Move();
```

对应的 C 代码如下:

```
typedef struct _IMovable
{
    Metadata* metaInfo;
    void (*Move)(_IMovable*); // C#: void Move()
} IMovable;

static int IMovableTypeId = 6;
```

```c
IMovable * NewIMovable()
{
    IMovable * result = (IMovable *)malloc(sizeof(IMovable));

    result->metaInfo = NewMetadata(result, IMovableTypeId, NULL, NULL);
    result->Move = NULL;

    return result;
}

typedef struct _Vehicle
{
    // basic info
    Metadata * metaInfo;
    // fields

    // methods

}Vehicle;

// type id
static int VehicleTypeId = 4;

// method declarations
void Move_Vehicle_IMovable_interface(IMovable * pIMovable);
void Move_Vehicle_IMovable(Vehicle * obj);

// the new method
Vehicle * NewVehicle()
{
    // init interfaces
    IMovable * pIMovable = NewIMovable();
    pIMovable->Move = Move_Vehicle_IMovable_interface;
    LinkNode * node = NewLinkNode(pIMovable->metaInfo);

    //alloc memory
    Vehicle * obj = (Vehicle *)malloc(sizeof(Vehicle));
    // initialize basic info
    obj->metaInfo = NewMetadata(obj, // pThis
        VehicleTypeId, // type id
        NULL, // base type
```

```
        node); // node of first interface.
    // initialize fields

    // initialize methods

    returnobj;
}
//the 'void Move();' method for class
void Move_Vehicle_IMovable(Vehicle * obj)
{
    printf("Vehicle.Move()\n");
}
//the 'void Move();' method for interface
void Move_Vehicle_IMovable_interface(IMovable * pIMovable)
{
    Vehicle * obj = (Vehicle * )Convert2Type(pIMovable->metaInfo, VehicleTypeId);
    Move_Vehicle_IMovable(obj);
}
```

由 Vehicle 类型、Metadata 和 interface 链表构成的描述 interface 的组织结构如图 B-4 所示。

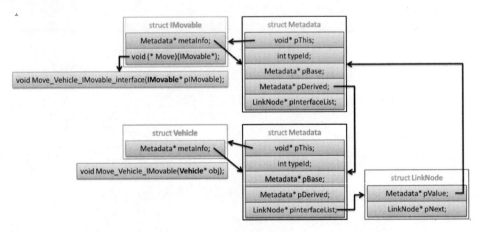

图 B-4 继承和实现了一个 interface 的 Vehicle 类型的对象

对应的 C 版代码的使用方式如下：

```
// C#:Vehicle obj = new Vehicle();
Vehicle* obj = NewVehicle();
// C#:obj.Move();
Move_Vehicle_IMovable(obj);
// C#:IMovable interfaceObj = obj as IMovable;
IMovable* interfaceObj = (IMovable*)Convert2Type(obj->metaInfo,IMovableTypeId);
// C#:interfaceObj.Move();
Move_Vehicle_IMovable_interface(interfaceObj);
```

有了 interface 以后，实际上就实现了"多继承"。一个 interface 类型可以被多个 class 类型继承，所以 C 版代码里，interface 里的函数指针类型的第一个参数只能是 interface 本身。即一个 class 类型 A，实现了 interface 类型 IX 里的方法，就意味着这个方法的第一个参数类型变成了 IX 的指针（而不再是 A 的指针）。

基于这样的 C 版设计，在调用接口函数时，一定会发生类型转换（调用 Convert2Type 函数）。例如，类型 Vehicle 实现了接口 IMovable，那么，Vehicle 就要为 IMovable 的各个函数分别创建声明完全相同的函数 Move_Vehicle_IMovable_interface(IMovable* pIMovable)，这些函数通过 Convert2Type 得到需要的类型指针 Vehicle*，再调用 Vehicle 内部的函数 Move_Vehicle_IMovable(Vehicle* obj)。

B.2.5 虚函数、接口与多次继承

如果接口的方法在基类里标记为 virtual，并且在子类和孙类里都 override 了，会怎么样？

```
interface IMovable
{
    void Move();
}
class Vehicle : IMovable
{
    public virtual void Move() // virtual method & interface method
    {
        Console.WriteLine("Vehicle.Move()");
    }
}

class Car :Vehicle
{
    public override void Move() // override method & interface method
    {
        Console.WriteLine("Car.Move()");
```

```
    }
}

class BMW : Car
{
    public override void Move() // override method & interface method
    {
        Console.WriteLine("BMW.Move()");
    }
}
```

对于这样的情况,下面的情形会输出什么?

```
BMW obj = new BMW();
VehiclebaseObj = obj;
IMovable interfaceObj = obj;
obj.Move();
baseObj.Move();
interfaceObj.Move();
```

答案是三行"BMW.Move()",即全部调用了最后 override 的方法。也就是说,转换出来的 C 代码中,除了基类的 virtual 方法的函数指针外,接口对象的函数指针也要修改为指向最后 override 的函数。从这个角度看,面向对象中的基类和接口的本质是相同的,其区别在于基类只能有 1 个,接口可以有多个。换句话说,可以认为基类是一个特别的接口。

B.2.6 其 他

像 public、protected、private 和 abstract 这样的关键字,用 C 是无法实现的,他们是面向对象语言在编译器进行语义分析时进行处理的。如果不符合规定(例如在类型外调用了标记为 private 的字段),编译器就会报错,停止生成代码,仅此而已。

最后强调一下容易混淆的地方。一个 class A 中的 virtual 方法,在 class B 中可以 override,然后还可以继续在 class C 中 override,其函数指针指向最后 override 的方法。classA 中的 virtual 方法,如果恰好也是 interface IX 中的方法,则 IX 中的函数指针也要指向最后 override 的方法。

B.3 用面向对象的方式描述 OpenGL 可以吗

CSharpGL 的目标是用面向对象的方式表达 OpenGL,因为 OpenGL 本身就是用 C 语言呈现的一个本质上面向对象的图形库。在 CSharpGL 中涉及到的 OpenGL 中隐含的主要类型如图 B-5 所示。

图 B-5 中,由 OpenGL 和一些最基本的数据结构构成基础,由这些基础隐含的

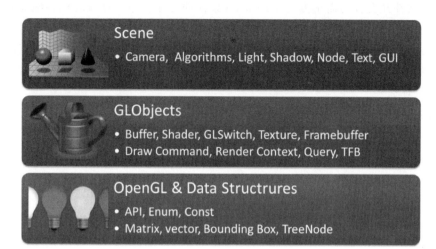

图 B-5 OpenGL 中隐含的主要类型

对象构成 OpenGL 的各个零件,由这些零件的各种组合形式构成了描述三维场景所需的设施。

B.3.1 OpenGL & Data Structures

OpenGL 呈现的形式是一些函数声明,以及这些函数声明使用的枚举类型或常量。OpenGL 中常用的数据结构包括矩阵(Matrix)、向量(vector)、包围盒(Bounding Box)、树节点(TreeNode)等。

B.3.2 GLObjects

OpenGL 函数以 C 语言的形式表达了面向对象的设计理念。这些常用的类型包括缓存(Buffer)、着色器(Shader)、状态开关(GLSwitch)、纹理(Texture)、帧缓存(Framebuffer)、渲染命令(Draw Command)、渲染上下文(Render Context)、查询对象(Query)和 Transform Feedback Object(TFB)等。

B.3.3 Scene

描述一个三维场景(Scene)所需的设施包括摄像机(Camera)、各种算法(Algorithms)、光源(Light)、阴影(Shadow)、节点(Node)、文字(Text)、图形用户界面(GUI)等。

B.4 关于面向对象程序设计的发散思考

问:如何学习设计模式?

答:先有面向对象思想,后有设计模式。先领会面向对象思想,再在实践中学设计模式。

问:面向对象程序设计在实际应用中的最大缺陷是什么?

答:软件开发者与需求方互不理解,导致面向对象程序设计没有机会施展到实际应用上。

问:为什么要把C封装为面向对象语言?面向对象语言是如何从无到有的?最开始的那个人是如何设计出这样一套机制的?他之前没有面向对象的任何概念,他的思路是什么?

答:只有对编程的热爱能给出这个答案。

附录C 解析简单的wavefront(*.obj)文件格式

学完本章之后,读者将能:
- 了解CSharpGL附带的wavefront文件解析器的设计;
- 了解CSharpGL附带的wavefront文件解析器的使用。

C.1 wavefront文件格式

扩展名为"*.obj"的wavefront文件是一种描述三维模型的文本文件,其格式非常简单。这种文件以纯文本的形式存储了模型顶点的位置、法线、纹理坐标和材质使用等信息。

C.1.1 单行格式

wavefront文件的每一行,都有极其相似的格式。在wavefront文件中,每行的格式如下:

```
prefix parameter1 parameter2 parameter3 ...
```

其中,前缀(prefix)标识了这一行所存储信息的类型,参数(parameter)则是具体的数据。wavefront文件常见的前缀如表C-1所列。

表C-1 wavefront文件格式常见前缀

前缀	含义
v	本行描述一个顶点的坐标。前缀后跟着3个float数值,分别表示该顶点的X、Y、Z坐标值。
vt	本行描述一个顶点的纹理坐标。前缀后跟着2个float数值,分别表示此纹理坐标的U、V坐标值。
vn	本行描述一个顶点的法线向量。前缀后跟着3个float数值,分别表示该法向量的X、Y、Z分量值。
f	本行描述一个面(Face)。一个面实际上就是一个三角形图元或者四边形图元。此前缀行的参数格式参见后面小节的详细介绍。
usemtl	此前缀后只跟着一个参数。该参数指定了从此行之后到下一个以usemtl开头的行之间的所有面所使用的材质名称。该材质可以在此OBJ文件所附属的MTL文件中找到具体信息。
mtllib	此前缀后只跟着一个参数。该参数指定了此OBJ文件所使用的材质库文件(*.mtl)的文件路径。

C.1.2 文件结构

在一个 wavefront 文件中,首先有一些以 v、vt 或 vn 前缀开头的行指定了所有顶点的位置、纹理坐标和法线。然后再由一些以 f 前缀开头的行指定每一个三角形(或四边形)所对应的顶点的位置、纹理坐标和法线的索引。在顶点的位置、纹理坐标和法线的索引之间,使用符号"/"隔开。一个 f 行可能出现的格式如表 C-2 所列。

表 C-2　f 前缀的格式

f 前缀的格式	含义
f 1　2　3 (f v v v)	以第 1、2、3 号顶点(以 v 为前缀的行)组成一个三角形。
f 1/3　2/5　3/4 (f v/vt v/vt v/vt)	以第 1、2、3 号顶点组成一个三角形。第一个顶点的纹理坐标的索引值为 3,第二个顶点的纹理坐标的索引值为 5,第三个顶点的纹理坐标的索引值为 4。
f 1/3/4　2/5/6　3/4/2 (f v/vt/vn v/vt/vn v/vt/vn)	以第 1、2、3 号顶点组成一个三角形,其中第一个顶点的纹理坐标的索引值为 3,法线的索引值为 4;第二个顶点的纹理坐标的索引值为 5,法线的索引值为 6;第三个顶点的纹理坐标的索引值为 4,法线的索引值为 2。
f 1//4　2//6　3//2 (f v//vn v//vn v//vn)	以第 1、2、3 号顶点组成一个三角形,且忽略纹理坐标。其中第一个顶点的法线的索引值为 4;第二个顶点的法线的索引值为 6;第三个顶点的法线的索引值为 2。

注意,wavefront 文件中的索引值是从 1 开始计数的,这与 C 语言中从 0 开始计数不同。在渲染的时候应注意将从文件中读取的坐标值减去 1。另外,一个 wavefront 文件里可能有多个模型,每个模型都是由(若干顶点属性信息+若干面信息)这样的顺序描述的。

C.2　多遍解析算法

对于任何一个有准确描述的格式,最好的方式就是使用编译原理来为其编写一个强大的解析器。这不仅可以解析任何符合格式的文件,还可以指出不符合格式的文件的错误之处。但是我并未找到 wavefront 格式的完全且准确的格式文法(Grammar),因此不可能做出这样的解析器。所幸 wavefront 的格式十分简单,可以用其他方式来解析。我提供一个具有灵活性的多遍解析算法,多遍解析 wavefront 的算法思路如图 C-1 所示。

首先,逐行遍历文件,找到文件的统计信息,例如顶点的位置数、UV 坐标数、法线数或面数。如果位置数与 UV 坐标数或与法线数不相等,就说明此文件是有问题的,从而停止解析。然后,根据已知的统计信息,创建相应的定长数组,用于存放顶点

图 C-1 多遍解析算法

属性（位置、UV 坐标、法线等）和面信息。注意，wavefront 中的面可能是三角形，也可能是四边形。定长数组比动态数组的读写效率更高，且能够在 CSharpGL 中直接用于创建 VertexBuffer。

最后，将模型的中心平移到原点（0，0，0）上，这样在旋转模型时才能保证模型始终不偏移。平移的距离和方向构成了模型在 World Space 里平移的逆向量。如果 wavefront 文件中没有记录模型的法线，就需要手动计算。如果模型的面是四边形而用户需要三角形，就要把每个四边形拆分成 2 个三角形。

用面向对象的方式，可以将每个遍历动作视为一个对象，这些对象继承自同一个抽象基类，基类有一个抽象的解析函数，从而让每个遍历动作完成各自不同的解析功能，代码如下：

```csharp
public class ObjVNFParser // parsing vertexes' positions, normals and faces in a *.obj file.
{
    private readonly ObjParserBase[] parserList; // array of step parsers

    public ObjVNFParser(bool quad2triangle)
    {
        var generalityParser = new GeneralityParser(); // find statistics info
        var meshParser = new MeshParser(); // find positions
        var normalParser = new NormalParser(); // find normals
        var texCoordParser = new TexCoordParser(); // find UVs
        var locationParser = new LocationParser(); // move model to origin(0, 0, 0)
        if (quad2triangle) // if need to transform quad to triangle
        {
            this.parserList = new ObjParserBase[] { generalityParser, meshParser, normalParser, texCoordParser, new Quad2TriangleParser(), locationParser, };
        }
        else
```

```csharp
        {
            this.parserList = new ObjParserBase[] { generalityParser, meshParser,
normalParser, texCoordParser, locationParser, };
        }
    }
    // parse an *.obj file
    public ObjVNFResult Parse(string filename)
    {
        ObjVNFResult result = new ObjVNFResult();
        var context = new ObjVNFContext(filename); // contains all parsed info
        try
        {
            foreach (var item in this.parserList) // multipass parsing
            {
                item.Parse(context);
            }

            result.Mesh = context.Mesh;
        }
        catch (Exception ex)
        {
            result.Error = ex;
        }

        return result;
    }
}
```

注意,用户可以根据需要灵活选择使用哪些解析步骤。例如,如果明确知晓要解析的 wavefront 文件中没有四边形面,那么就不需要加入 Quad2TriangleParser 这一步骤。

C.3 如何使用

使用此解析器十分简单,只需创建一个解析器对象,调用其解析函数,代码如下:

```csharp
private void FormMain_Load(object sender, EventArgs e)
{
    string filename = "HanoiTower.obj"; // wavefront file name
    var parser = new ObjVNFParser(false); // create a parser
    ObjVNFResult result = parser.Parse(filename); // parse specified wavefront file
    if (result.Error != null) // if error exists, tip it
```

```
        {
            MessageBox.Show(result.Error.ToString());
        }
        else // create a node to display wavefront file model in CSharpGL
        {
            var node = ObjVNFNode.Create(result.Mesh);
            float max = node.ModelSize.max();
            node.Scale *= 7.0f / max; // scale model to appropriate size
            node.WorldPosition = new vec3(0, 0, 0); // discard where model is in file
        }
    }
```

读者可打开本书网络资料的项目 Demos\SimpleObjFile,查看加载 wavefront 模型的示例,如图 C-2 所示。

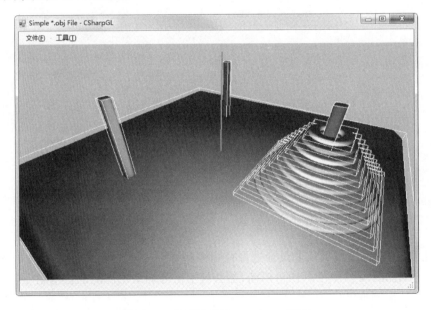

图 C-2 加载和渲染 wavefront 模型

附录 D 自制体数据的 2 种方法

学完本章之后,读者将能:
- 了解如何使用数学公式法自制体数据;
- 了解如何使用分层雕刻法自制体数据。

学习研究体渲染(Volume Rendering)的困难之一是难以找到合适的体数据。本章介绍 2 个自制体数据的方法,使用这两个方法可以获取符合测试需要的体数据,既可以保存为文件,又可以在每次使用时生成,不需要持续占用大量的空间。

首先要明确,本书涉及的体数据都是用三维纹理(3D Texture)表示的。为其填充的数据可以用一个数组表示。例如本书网络资料的 Demos\VolumeRendering.Raycast 项目中的体数据是用 byte[]这一数组类型填充的。也就是说,只需为一个数组填充合适的数值,即可得到体数据。为了让体数据自身带有颜色,本章用 Voxel[]这一数组类型描述体数据,其中的 Voxel 类型定义如下:

```
// voxel data in volume rendering
public struct Voxel
{
    public byte r; // red component
    public byte g; // green component
    public byte b; // blue component

    public Voxel(byte r, byte g, byte b)
    {
        this.r = r; this.g = g; this.b = b;
    }

    public static Voxel operator +(Voxel x, Voxel y)
    {
        byte r = (byte)(x.r + y.r);
        byte g = (byte)(x.g + y.g);
        byte b = (byte)(x.b + y.b);
        return new Voxel(r, g, b);
    }
}
```

Voxel 即"体素",是描述体数据在立体空间某处的数据,类似描述图片的"像素"。

D.1 数学公式法

使用简单的解析几何公式即可在体数据中雕刻出漂亮的形体。如式(D-1)所示。

$$\left.\begin{array}{l} Ax + By + Cz + D = 0 \\ \dfrac{x^2}{a^2} + \dfrac{y^2}{b^2} + \dfrac{z^2}{c^2} + D = 0 \\ \dfrac{x^2}{a^2} - \dfrac{y^2}{b^2} - z = 0 \end{array}\right\}$$ 式 D-1 平面(上) 椭球体(中) 马鞍面(下)

这些公式可以用 $f(x,y,z)=0$ 来统一表示。对于一个宽、高、深度分别为

width、height和depth,用于描述体数据的Voxel[]类型来说,其元素数量应为"width * height * depth"。可以将它视为一个由width、height和depth确定了的长方体。因此可以用下述伪代码来为其雕刻出解析几何图形:

```csharp
static Voxel[] GetData(int width, int height, int depth)
{
    var result = new Voxel[width * height * depth];
    for (int i = 0; i < width; i++)
    {
        for (int j = 0; j < height; j++)
        {
            for (int k = 0; k < depth; k++)
            {
                var index = i * height * depth + j * depth + k;
                double x = (i - width / 2.0);
                double y = (i - width / 2.0);
                double z = (j - height / 2.0);
                doubledelta = f(x, y, z);
                if (0 <= delta && delta < 0.1) // if delta is close enough
                {
                    result[index] += new Voxel(...); // this is where something exists
                }
            }
        }
    }

    return result;
}
```

此代码的核心是,将数组视为一个中心在原点的长方体,将根据描述数组的索引{i, j, k}得到的坐标(i-width / 2.0, j-height / 2.0, k-depth / 2.0)视为长方体内的一个点。如果此点满足公式$f(x, y, z)=0$(或者$f(x, y, z)$不超过指定阈值),那么就在数组的此索引位置加上指定的值(此值代表其将显示的颜色)。

为了在体数据中添加任意多个解析几何图形的痕迹,可以将每个解析几何公式封装起来,代码如下:

```csharp
//a plane: Ax + By + Cz + D = 0
class ShapingPlane : ShapingBase
{
    // parameters of a plane
    double A;
    double B;
```

```csharp
        double C;
        double D;

        public ShapingPlane(double a, double b, double c, double d, int width, int height, int depth)
            : base(width, height, depth)
        {
            this.A = a; this.B = b; this.C = c; this.D = d;
        }

        // engrave some value at a specified index
        public override Voxel Shaping(int i, int j, int k)
        {
            double x = (i - width / 2.0);
            double y = (i - width / 2.0);
            double z = (j - height / 2.0);
            double delta = A * x + B * y + C * z + D;
            if (0 <= delta && delta < 0.1) // if delta is close enough
            {
                return new Voxel(0, byte.MaxValue, 0); // this is where something exists
            }
            else
            {
                return new Voxel();
            }
        }
    }

    // base type for shaping
    abstract class ShapingBase
    {
        // size of volume data
        protected int width;
        protected int height;
        protected int depth;
        // engrave some value at a specified index
        public abstract Voxel Shaping(int i, int j, int k);

        public ShapingBase(int width, int height, int depth)
        {
            this.width = width; this.height = height; this.depth = depth;
        }
    }
```

此时雕刻体数据的代码则如下：

```csharp
static Voxel[] GetData(int width, int height, int depth)
{
    list = new List<ShapingBase>(); // list of shapes.
    list.Add(new ShapingPlane(0, 1, 0, -2.0 * width / 5.0, width, height, depth));
    //add a plane
    list.Add(..); // add something else

    var result = new Voxel[width * height * depth];
    for (int i = 0; i < width; i++)
    {
        for (int j = 0; j < height; j++)
        {
            for (int k = 0; k < depth; k++)
            {
                var index = i * height * depth + j * depth + k;
                foreach (var item in list) // engrave all shapes in the list
                {
                    Voxel value = item.Shaping(i, j, k);
                    result[index] += value;
                }
            }
        }
    }

    return result;
}
```

使用这一方法，得到满意的体数据 Voxel[] 后，可以用下述伪代码将其封装为三维纹理对象：

```csharp
// Get 3D texture object from volume data
Texture InitVolumeTexture(Voxel[] data, int width, int height, int depth)
{
    var storage = new TexImage3D(GL_TEXTURE_3D, GL_RGB, width, height, depth, GL_RGB,
GL_UNSIGNED_BYTE, new ArrayDataProvider<Voxel>(data));
    var texture = new Texture(storage, new MipmapBuilder(),
        new TexParameteri(GL_TEXTURE_WRAP_R, GL_CLAMP_TO_EDGE),
        new TexParameteri(GL_TEXTURE_WRAP_S, GL_CLAMP_TO_EDGE),
        new TexParameteri(GL_TEXTURE_WRAP_T, GL_CLAMP_TO_EDGE),
        new TexParameteri(GL_TEXTURE_MIN_FILTER, GL_LINEAR_MIPMAP_LINEAR),
        new TexParameteri(GL_TEXTURE_MAG_FILTER, GL_LINEAR),
```

```
            new TexParameteri(GL_TEXTURE_BASE_LEVEL, 0),
            new TexParameteri(GL_TEXTURE_MAX_LEVEL, 4));
    texture.Initialize();

    return texture;
}
```

D.2 分层雕刻法

D.2.1 基本流程

分层雕刻法(Per-layered Engrave)是通过对 FrontToBackPeeling 算法进行改造而得的。在 FrontToBackPeeling 算法中,每次只渲染当前最靠近 Camera 的 Fragment,将其与之前已经渲染的 Fragment 进行混合操作,然后将此次渲染的 Fragment 指定为已渲染过。通过逐层地渲染实现了与顺序无关的半透明效果。这里将逐层的混合操作修改为逐层的雕刻体数据操作,即可得到所需的体数据。

关于 FrontToBackPeeling 算法请参考本书的 13.3 节。

在第一次渲染后,得到了第一层 Fragment 构成的画面,可将其导出为一个 PNG 格式的图片。图片上的每个像素包含 RGBA 共 4 个分量,描述了三维纹理中的一个点。像素在图片中的位置即为此点的 XY 坐标,像素的颜色值 RGB 即为此点的颜色值,像素的 A 分量即为此点的 Z 坐标。第一次渲染的 Shader 代码如下:

```
// vertex shader
#version 330 core

in vec3 inPosition; //object space vertex position

uniform mat4 mvpMat; //combined modelview projection matrix

void main()
{
    //get the Clip Space vertex position
    gl_Position = mvpMat * vec4(inPosition, 1);
}

// ---------------------------------------------------------
// fragment shader
#version 330 core

out vec4 outColor; //output fragment color

uniform vec4 color;//color uniform
```

```glsl
void main()
{
    outColor = vec4(color.rgb, gl_FragCoord.z);
}
```

此代码与FrontToBackPeeling算法中的第一次渲染代码的区别仅在于,用gl_FragCoord.z(内置变量,表示深度值)填充了Fragment的Alpha分量,这样就可以让随后导出的PNG图片的A分量描述Z坐标。根据PNG图片雕刻体数据的伪代码如下:

```csharp
// engrave aslice(png) into volume data
void GetVolumeData(Voxel[] volumeData, Bitmap png, Color background, int width, int height, int depth)
{
    for (int w = 0; w < width; w++)
    {
        for (int h = 0; h < height; h++)
        {
            Color c = png.GetPixel(w, h);
            int d = (int)((double)depth * (double)c.A / 256.0);
            int index = w * height * depth + h * depth + d;
            // if this pixel is not on edge or background, we engrave it into volume data
            if (c.A != 0 && c.A != byte.MaxValue &&
                (c.R != background.R || c.G != background.G || c.B != background.B))
            {
                volumeData[index] += new Voxel(c.R, c.G, c.B);
            }
        }
    }
}
```

在之后的每次Peeling渲染后,都会得到一层Fragment构成的数据,均可以将其导出为一个PNG格式的图片。像第一次渲染得到的PNG图片一样,可以据此继续雕刻体数据,直至所有Peeling渲染都结束。每次Peeling操作的Shader代码如下:

```glsl
// vertex shader
#version 330 core

in vec3 inPosition; //object space vertex position

uniform mat4 mvpMat; //combined modelview projection matrix
```

```
void main()
{
    //get the Clip Space vertex position
    gl_Position = mvpMat * vec4(inPosition, 1);
}

// -------------------------------------------------
// fragment shader
#version 330 core

out vec4 outColor; //fragment shader output

uniform vec4 color; //color
uniform sampler2DRect depthTexture; //depth texture

void main()
{
    // read the depth value from the depth texture
    float depth = texture(depthTexture, gl_FragCoord.xy).r;

    // compare the current fragment depth with the depth in the depth texture
    // If it is less, discard the current fragment
    if(gl_FragCoord.z <= depth) discard;

    // Otherwise set the given color uniform as the final output
    outColor = vec4(color.rgb, **gl_FragCoord.z** );
}
```

此代码与 FrontToBackPeeling 算法中的 Peeling 代码的区别仅在于用 gl_FragCoord.z(内置变量,表示深度值)填充了 Fragment 的 Alpha 分量,这样就可以让随后导出的 PNG 图片的 A 分量描述 Z 坐标。

简单地说,分层雕刻法将 FrontToBackPeeling 算法中的 Blend 操作变为雕刻操作,从而将三维场景转换为体数据。将本书使用的汉诺塔模型雕刻到三维纹理从而进行体渲染,如图 D-1 所示。

D.2.2 注意事项

首先,在每次渲染时,应使用正交投影方式(glOrtho(..))的 Camera。这样才能保证三维场景的远处和近处被同等地雕刻到体数据中。渲染时的画布大小必须与体数据的 Width 和 Height 值相同。如果不同,就要对每层得到的 PNG 图片进行缩放,以匹配体数据的大小。在三维场景中,模型边缘的四周的深度值会剧烈变化。缩放操作将导致模型边缘处的深度值被插值处理(在剧烈变化的深度值中取加权平均

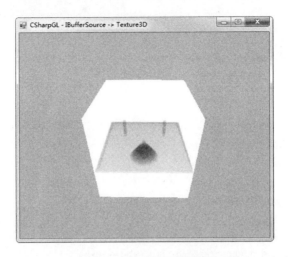

图 D-1 汉诺塔的体渲染

值),从而与其应有的深度值产生巨大的误差,这将导致模型周围被雕刻到体数据的错误位置上,然后通过 glViewport(..) 函数调节视口的大小,使之与画布大小相同。

最后,要注意调节 Camera 的参数,尽量使之恰好包围要渲染的目标。如果 Camera 的 Near 和 Far 设置太大,会降低 PNG 图片 Alpha 分量的精确度。一个简单的方法是用中心位于原点的模型的大小 vec3 size; 来设置 Camera 的参数,代码如下:

```
void SetupCamera(ICamera camera, vec3 size)
{
    IOrthoViewCamera c = camera;
    c.Left = -size.x / 2; c.Right = size.x / 2;
    c.Bottom = -size.y / 2; c.Top = size.y / 2;
    c.Near = -size.z / 2; c.Far = size.z / 2;
}
```

按此代码设置 Camera 的参数,可以使 Camera 恰好仅仅观察到目标的范围。

D.2.3 Compute Shader 雕刻

在用 PNG 图片雕刻体数据的过程中,对各个像素的雕刻操作都是互不相关的,输入数据来自二维纹理对象,输出数据是三维纹理对象,这完全符合 Compute Shader 的应用情景。因此可以用 Compute Shader 来代替客户端进行雕刻,代码如下:

```glsl
// compute shader
#version 430 core

layout (local_size_x = 256) in; // width of future 3D Texture is hard coded 256

// image that representing a layer
layout (rgba32f, binding = 0) uniform image2D imgLayer;
// volume data
layout (std430, binding = 1) buffer outputBuffer {
    uint outBuffer[];
};

//uniform int width = 256; not needed in this shader
uniform int height = 256;
uniform int depth = 256;

void main()
{
    // win 'for (int w = 0; w < Width; w++)' in client code
    uint w = gl_GlobalInvocationID.x; // or uint w = gl_LocalInvocationID.x

    // h in 'for (int h = 0; h < Height; h++)' in client code
    uint h = gl_GlobalInvocationID.y; // or uint h = gl_WorkGroupID.y

    // color in 'Color color = bitmap.GetPixel(w, h);' in client code
    vec4 color = imageLoad(imgLayer, ivec2(w, h));

    uint d = uint(depth * color.a);
    if (d == depth) { d = depth - 1; }
    uint index = uint(w * height * depth + h * depth + d);
    outBuffer[index] = packUnorm4x8(color); // packed color
}
```

注意,这里为了节省需要的空间,用 uint 来记录压缩了的 RGBA 颜色值。这与介绍顺序无关的透明小节 13.2.2 时使用的方法相同。

使用 Compute Shader 进行雕刻的客户端代码如下:

```
// engrave volume data with compute shader
    void ComputeShaderEngrave(Texture texInput, VertexBuffer outBuffer, int width, int height)
    {
        uint inputUnit = 0, outputUnit = 1;
```

```
        ShaderProgram computeProgram = this.engraveComp;

        // Activate the compute program and bind the output texture image
        computeProgram.Bind();

        // bind texture's first image as a layer to be engraved
        glBindImageTexture(inputUnit, texInput.Id, 0, false, 0, GL_READ_WRITE, GL_RGBA32F);
        // engrave into this buffer object
        glBindBufferBase(GL_SHADER_STORAGE_BUFFER, outputUnit, outBuffer.BufferId);

        glDispatchCompute(1, (uint)height, 1); // execute compute shader
        glMemoryBarrier(GL_SHADER_IMAGE_ACCESS_BARRIER_BIT);

        glBindImageTexture(inputUnit, 0, 0, false, 0, GL_READ_WRITE, GL_RGBA32F);

        computeProgram.Unbind();
    }
```

使用 Compute Shader 能够提升雕刻效率，但是在不支持 Compute Shader 的环境下，用客户端方式进行雕刻也是一个选项。Compute Shader 雕刻后得到了一个 VertexBuffer 对象，可以用 glMapbuffer(..) 的方式获取其中的内容，代码如下：

```
Voxel[] GetVolumeData(VertexBuffer buffer, int width, int height, int depth)
{
    var volumeData = newVoxel[width * height * depth]; ;
    var array = (uint*)buffer.MapBuffer(MapBufferAccess.ReadWrite);
    for (int i = 0; i < volumeData.Length; i++)
    {
        vec4 color = glm.unpackUnorm4x8(array[i]);
        volumeData[i] = new Voxel(
            (byte)(color.x * 255.0),
            (byte)(color.y * 255.0),
            (byte)(color.z * 255.0));
    }
    buffer.UnmapBuffer();

    return volumeData;
}
```

在服务器端进行压缩的 RGBA 格式（uint 类型）可以在客户端进行解压，得到由 vec4 表示的 RGBA 值。